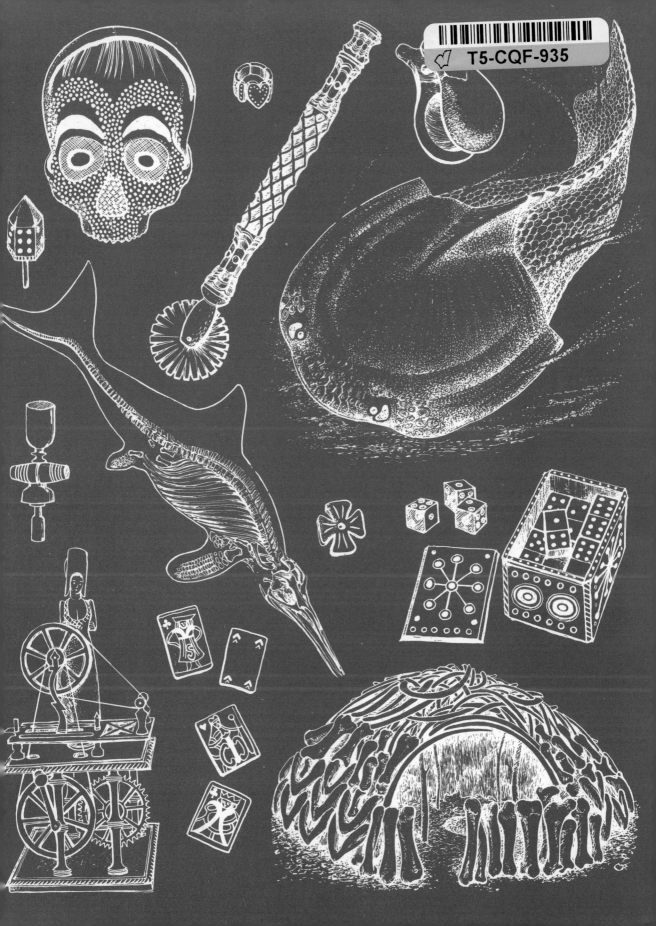

Bare Bones

An exploration in art and science

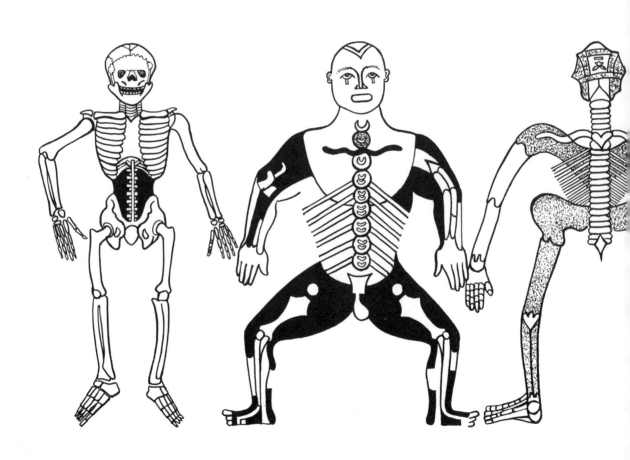

Beverly Halstead
Jennifer Middleton

Bare Bones

An exploration in art and science

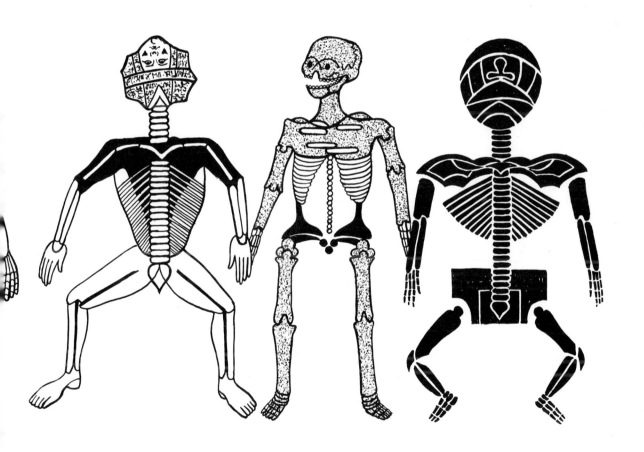

University of Toronto Press

OLIVER & BOYD
Tweeddale Court
14 High Street
Edinburgh EH1 1YL

A Division of Longman Group Limited

First published 1972
First published 1973 in Canada
and the United States by
University of Toronto Press
Toronto and Buffalo
ISBN 0-8020-1971-4
ISBN (Microfiche) 0-8020-0314-1

Set in 'Monophoto' 11/13 Ehrhardt
by Photoprint Plates Limited, Rayleigh, Essex
Printed in Great Britain by
T. & A. Constable Limited, Edinburgh

Contents

But who knows the fate of his bones, or how often he is to be buried? Who hath the Oracle of his ashes, or whither they are to be scattered?

. . . . .

To be knav'd out of our graves, to have our sculs made drinking-bowls, and our bones turned into Pipes, to delight and sport our Enemies, are Tragicall abominations escaped in burning Burials.

. . . . .

Now since these dead bones have already out-lasted the living ones of *Methuselah*, and in a yard under ground, and thin walls of clay, out-worn all the strong and specious buildings above it; and quietly rested under the drums and tramplings of three conquests; what Prince can promise such diuturnity unto his Reliques, or might not gladly say,

Sic ego componi versus in ossa velim.

Time which antiquates Antiquities, and hath an art to make dust of all things, hath yet spared these *minor* Monuments.

From Sir Thomas Browne:
Hydriotaphia: Urne Buriall, 1658

Preface

Bones are handled in one way or another by orthopaedic surgeons, archaeologists, antique collectors and artists. Their approaches are vastly different although the subject is the same. Each group has evolved its own jargon which serves to isolate it from the others. We have written this book for the layman and even though any one expert may find a section dealing with his own speciality, most of the book will be outside his field—in it he too will be a layman.

This is a book about bones but not, we trust, a dry book. We have looked at bones from many different angles, we discuss where they came from, why we have them in the first place, how they work and what goes wrong with them. Then we go on to outline the uses to which man has put them and briefly describe a bone-based culture. The last part of the book is about bones in art, their use as the medium of expression and finally their actual portrayal, tracing the changing attitudes towards them.

Our account does not claim to be comprehensive; it is simply our personal choice. We enjoyed producing it; we hope the reader will find it fun.

Introduction 1

Inside every one of us there is a skeleton of which we are little aware, unless we are unfortunate enough to fracture a rib or break a leg. Without a skeleton we would be unable to stand up or move about, and without it life as we know it would not be possible. In spite of this, our skeleton is something we do not like to contemplate. It symbolises death and the reason is fairly obvious. For the bare bones to be visible, all the flesh has to go and this naturally enough involves the death of the individual concerned. The transformation of a face to a skull has elements of real horror. This was suitably exploited by Alfred Hitchcock in his film *Psycho*; at the end of the film the young woman, pursued by the murderer, escapes into the cellar, sees the murderer's old mother sitting in her chair and turns it to be met by a grinning skull atop a dressed corpse. Horror indeed!

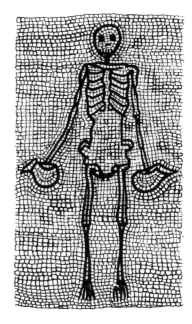

Roman mosaic

A skeleton was so evocative of death that for many centuries it seems no one looked at one very closely. They are generally represented in crude terms—to our modern eyes amusing, if not exactly endearing. During the Renaissance, beginning with the great anatomist Vesalius, the skeleton was examined and portrayed with accuracy. It was a shock to all true believers to discover that man and woman had the same number of ribs. The Church did not take kindly to his work, and like all pioneers he was persecuted for his pains.

Today, through the advent of X-rays, we are fortunate in being able to examine our skeletons without waiting for death. Most people are familiar with the grey strips which appear on the transparency the dentist shows his patients and from which he confidently identifies all sorts of dire conditions requiring appropriate treatment. Until the shop X-ray machines were banned, many children had seen the bones of their feet when they went to get a new pair of shoes. Periodically we are encouraged to take advantage of the Mass X-ray Units, which tell us whether smoking has damaged our lungs. These pictures are all rather anonymous—a rib cage is a rib cage is a rib cage.

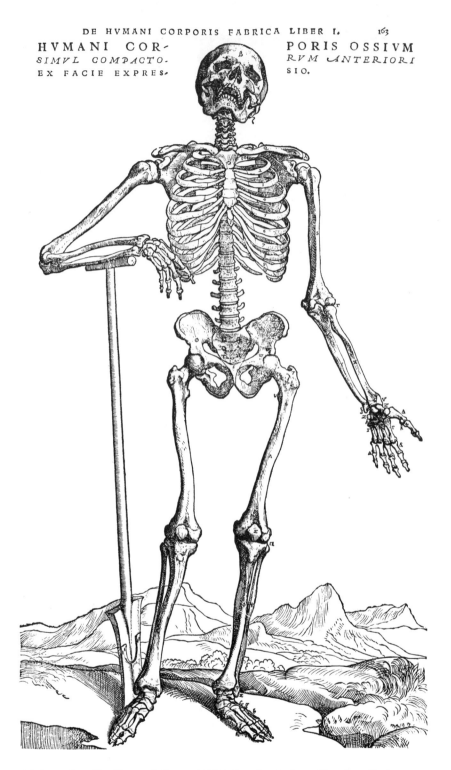

HVMANI COR-
SIMVL COMPACTO-
EX FACIE EXPRES,

PORIS OSSIVM
RVM ANTERIORI
SIO.

Skeleton from De Humani Corporis Fabrica, *by Andreas Vesalius, Basle, 1543.*

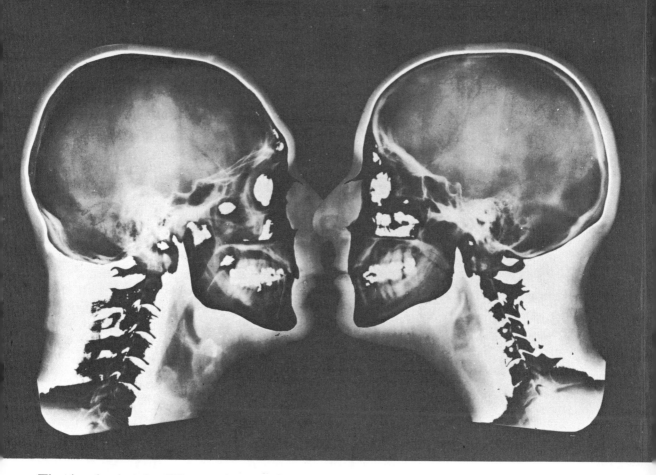

The situation is quite different when it comes to viewing radiographs of one's own head, or so at least we discovered. Our photograph shows both the bones and the associated soft tissues of our heads and necks. Unfortunately hair is not registered, so that we appear bald (we are not!). The first sight of the inside of our heads was an eerie experience. To the chagrin of certain protagonists of woman's liberation we could not help but be struck by the differences. For example, the bony receptacle for the pituitary gland, which controls all the other endocrine glands and the production of their respective hormones, was considerably larger in the male. The male face was more heavily ossified and the neck vertebrae more massive.

A feature which occasioned most surprise was the discrepancy between the flesh of the face and the underlying bone. The distances from the skeleton to the skin of the nose, lips and chin were much greater than we had imagined. When the heads of prehistoric man are reconstructed from the skulls, taking this fact into account, they suddenly appear more human looking, as can be seen clearly in our new reconstruction of Rhodesian man, illustrated at the beginning of the chapter 'Bones and Disease'.

Excellent reconstruction of faces from skulls were made by the late Professor Gerasimov of Moscow. He studied the delicate moulding which the muscles of the face make on bones. Perhaps his most famous case was that of Ivan the Terrible. His restorations were of such accuracy that he was frequently employed by the Soviet police for restoring the faces of disinterred skulls; the results were then checked in police files. This work emphasises the fact that the shapes of bones are modified by the action of the muscles. To a considerable extent, and certainly in evolutionary terms, the form of the skeleton reflects our function. For example, people who squat frequently on their haunches develop small squatting facets at the base of their shin bones. The Patagonians who hunt with the bolas—three balls on the end of thongs which they whirl round their heads and then hurl at the legs of running animals—develop a spiral grooving on their upper arm bones.

Our skeletons provide us with our internal support. But this alone is not enough, unless we are sedentary organisms. We have to move about to some extent, hence the skeleton must be jointed, and by virtue of this we can walk, write and fight.

The other major role of our skeleton is protection. The most delicate organ in the body is the brain—if the skull is staved in, the brain splashes. The cranium is so important that many zoologists do not speak of vertebrates but rather of 'craniates'. The ribs and backbone also have certain protective functions. The roles of support and protection are reasonably obvious but the skeleton has other, more hidden, tasks. It acts as a chemical store, mainly for the element calcium, the vital need for which is currently being exploited by the advertising campaigns of the Milk Marketing Board. Growing children require calcium for bone formation, and until the introduction of free school milk many children from poor families suffered from rickets. As a consequence of insufficient calcium, vitamin D and sunlight, the limb bones were weakly developed and were unable to support the child's weight properly. The bones buckled, resulting in a life-long disability. Free school milk eliminated this deficiency disease in Great Britain and it is sad to have to record that the present British government has seen fit not only to withdraw it but even refuses to allow local authorities to continue its supply.

We can attribute our success on earth to our bony skeleton, yet we know that we are descended from animals lacking any bony internal support. The questions come to mind naturally enough: where did we get this skeleton from in the first instance, and why? This is the topic we shall discuss in the next chapter.

The why 2
and wherefore
of bone

To answer the question 'Where did bones come from, and why?', we have to return to a period five hundred million years ago, a time when there were no backboned animals as we know them today. Living in the sea, however, were a group of primitive fish-like animals—the ostracoderms. Unlike true fishes these animals did

Siberian ostracoderms from 370 million years ago.

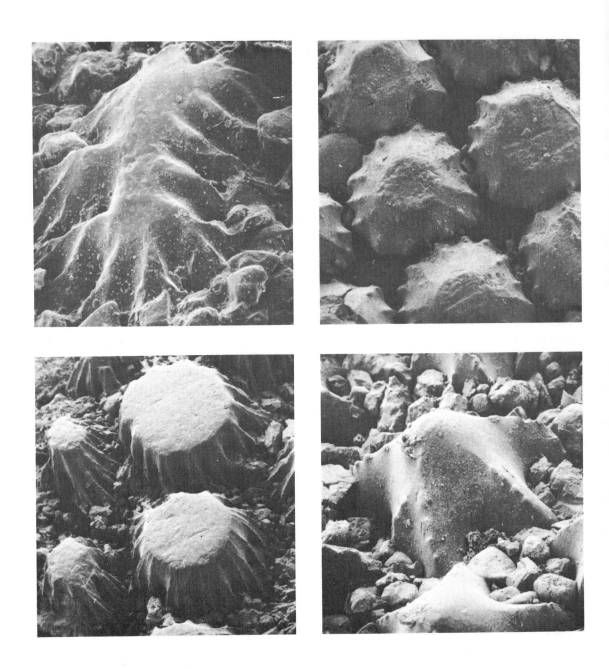

Details of ostracoderm armour photographed under the electron scanning microscope.

not have jaws. Naturally they had mouths, but they were incapable of biting anything. Their lives were spent sucking mud off the sea floor—a not very auspicious start to the lineage which gave rise to ourselves. From the remains of these creatures, preserved in ancient rocks we can discover many things about their way of life. These animals did not have a bony skeleton on their inside, but instead had an armour of bony plates on their outside. The internal skeleton consisted of gristle or cartilage, not bone.

We know that bone first appeared in the skin of the earliest vertebrates and its function must have been quite different from what it is today in ourselves. The outer part of this armour was thrown up into tiny bumps or tubercles which, when microscopically examined in section, are seen to be composed of dentine, the main substance of teeth and ivory. Again, the tissue of these small tooth-like tubercles must have had a different function from its present one. It has been suggested that it was concerned primarily with sensitivity, as it was the main barrier between the animal and its environment. It is our common experience that this facility is retained, frequently to an excruciating degree! The retention of this sensitivity in what should be simply dead and stone-like objects, intended merely for chewing our food, has been termed one of evolution's practical jokes!

Bone and dentine are similar mineralised tissues composed of a fibrous protein, collagen, on which are deposited tiny hexagonal-prismatic crystals of the mineral apatite (calcium phosphate).

Dentine is the material of ivory and can be distinguished from bone because the latter is penetrated by fine blood vessels which appear as minute dots, or short fine lines, enabling one to tell whether an 'ivory' being offered for sale is actually bone!

Mineral deposits can be very readily laid down in many parts of the body. For example, when teeth are scaled by the dentist, he actually chips off a deposit (calculus) from the inner surface of the lower front teeth. This calculus is made up of various forms of crystalline calcium phosphate, together with other substances, including bacteria.

In more confined spaces in the body, different mineral formations can occur. In the gall bladder, for instance, stones of cholesterol, calcium salts and bile pigments are fairly common; more so in women than men and especially in those of ample proportions. Usually gallstones are irregular and facetted, the facetting being due to constant friction and pressure within the gall bladder. Gallstones take a long time to grow and the specimen illustrated was from a 45 year-old man who had a feeling of discomfort in his right abdomen for only the previous nine months.

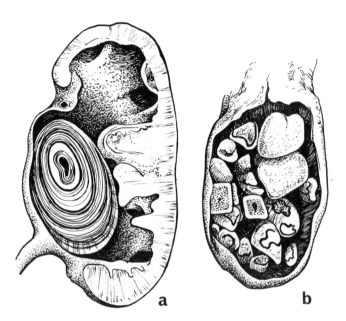

(a) Kidney blocked by growth of large stone; (b) gall bladder crammed full of facetted stones.

a

b

Seven tetrahedral stones removed from a man's bladder.

Stones in the kidney are formed from the crystallised constituents of urine. There is a need to drink reasonably copious amounts of liquids to ensure the urine does not become too saturated. Renal stones are composed of uric acid and calcium and other urates, with occasionally some other salts. The large calculus illustrated was from a 40 year-old man who had been struck in his right loin at the age of 20. There had been pain, and blood was present in his urine and although it subsided he had had painless haematuria (blood in the urine) since. The laminated stone measured 3 by 2 inches and blocked the outlet of the kidney. In the apparently greatly enlarged kidney, the actual amounts of functioning kidney tissue were reduced to a narrow outer zone.

Stones in the bladder, however, provide some of the more bizarre anecdotes of the medical profession. Vesical calculī, as they are termed, may be single or multiple and may even look like coarse sand. These last are painfully voided during micturition. One of the most attractive examples is from a 25 year-old Nigerian man who suffered pain and bleeding at the end of micturition; from his bladder seven compacted tetrahedral stones were removed.

Other calculi we have examined were those removed from Napoleon III of France and Leopold I of Belgium. These cases formed naturally, but the presence of any foreign object in the bladder provides a nucleus around which mineral salts may be deposited. A bilobed stone, extracted from a 29 year-old man, contained an open tie-pin. One specimen was deposited around a piece of chewing-gum, which had been inserted into the penis five years earlier. A boy of 17 had introduced an uncooked pea into his urethra 14 months previously and in this short period it had formed a stone 2 inches in diameter; when it was sectioned, the pea was still intact at the centre of the stone.

For these minerals to be laid down, they must be in sufficient concentration allowing them to precipitate out of solution; and there must be some kind of seeding nucleus, such as a tie-pin or chewing-gum! The body appears to have the facility for calcifying objects as opposed to merely coating them. A foetus which dies, but instead of being aborted is retained within the uterus, will become impregnated with calcium phosphate. Sir William Cheselden, surgeon to Queen Anne, described in 1733 such a stone child (or lithopedion) that had been in a woman for 26 years. Recently there was an account in the newspapers of an Italian woman who had carried a child for 37 years without even knowing it. This must also have been a lithopedion. In some instances, as in the one we have illustrated, they look like primitive carvings.

Lithopedion or stone child (after Selye).

Although calcium phosphate can be laid down in many sites of the body, in the first vertebrates it was confined to the skin. It was not in the form of stones but rather an organised spongy meshwork, a delicate three-dimensional scaffolding. Moreover, it was not purely mineral matter but was partially made of the fibrous protein collagen. Contrary to all expectations, this organic matter has survived for five hundred million years and the amino-acids, the building blocks of proteins, can be extracted from the fossils by modern biochemical techniques.

Collagen makes up about one third of all the protein of the body, it is present in all connective tissues, is the basic material of tendons and ligaments, and provides the organic matrix of bones and teeth. Experiments in the laboratory have shown that when collagen is introduced into a solution of calcium phosphate, this mineral will readily precipitate. In fact, collagen is the ideal seed bed on which crystals of apatite can form. It is believed that the unique structure of the collagen molecule allows the crystals to form on it. The greatest concentration of collagen is found in the skin: hence it should be no surprise to learn that it was in skin that the first bone developed.

Many primitive animals still have bones in their skins. The patterning of crocodile skin which makes it much sought after for handbags and shoes, and threatens the survival of the crocodile, is the result of its being moulded on the sculptured bony plates beneath the skin surface. These bony scutes give the animal an almost impregnable armour. A more extreme example is provided by the tortoise whose carapace is a bony box open only at the front and back. Even this armour has a further outer covering of tortoiseshell, another type of fibrous protein called keratin, the material of which our nails and hair are composed.

In evolutionary history, bone is always primarily associated with the skin, and formation of bone in this part of the body is something that can always be reactivated. Such bone formation can be initiated very readily and can have unfortunate consequences for the people concerned. Where foreign materials are inserted in the body, as in reconstructive surgery of the breast, some bizarre results occur. A material that has been used in such breast prostheses is a synthetic sponge, which was considered especially suitable because it did not produce any tissue reaction. One of the most promising of these synthetic sponges was made of polyhydroxy-ethylmethacrylate (polyHEMA: Hydron), and in the late 1960s this material was used in Czechoslovakia in about a hundred breast prostheses. Dr George Winter of the Institute of Orthopaedics, London, has been implanting the same material into the skin of pigs and has discovered that, after some 60 days, these implants are converted into bone, the spaces in the sponge being filled with typical bone and the plastic material itself being impregnated with calcium phosphate. It has been noted that in these Czechoslovakian women there is a radio-opaque shadowing which suggests that some form of mineralisation or calcification is taking place. It is believed that one month of growth in the pig is equivalent to about one year in the human and so it is expected that, in time, these women will have large bony lumps. Naturally enough,

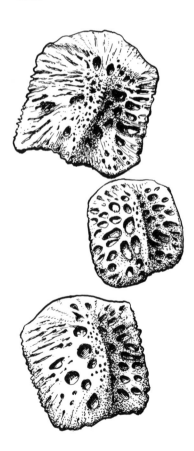

Bony scutes underlying crocodile skin.

following the work of Dr Winter, this material is no longer being used. What this cautionary tale emphasises is the ease with which bone can be formed in the skin, be it in the back of a pig or the breast of a woman. Although skin retains the ability to produce bone, this does not tell us why it has this potentiality in the first instance.

It is evident that the presence of collagen is an important factor. But in spite of this, there had to be some mechanism for concentrating calcium phosphate to enable mineralisation to take place. Such a mechanism was discovered a few years ago by Drs John Greenspan and Irving Shapiro of the Royal Dental Hospital, London. They recognised that the answer was to be found in the structure of the cell. The cells of animals, and plants for that matter, contain many structures which can only be seen with the aid of the electron microscope. These miniature cellular organs, or organelles, each have their own allotted tasks to perform. One of the key organelles is called the mitochondrion, the power-house of the cell, within which take place all the energy-exchange mechanisms supporting life. However, the mitochondrion possesses another property; it has the most amazing affinity for calcium and phosphate ions. It soaks them up, moreover, in the same proportions as they occur in the mineral apatite. The mitochondrion collects these ions, after which it spews them out.

After many years of searching, on the part of scientists, for the basic mechanism of biological mineralisation, Greenspan and Shapiro finally laid their hands on the critical clue. Their work has provided us with the most likely mechanism explaining the origins of bone; but it has not suggested the reason *why*.

For the answer to this, it is necessary to return to the ostracoderms and to consider the conditions under which they lived 500 million years ago. These creatures inhabited the sea and were faced with peculiarly marine chemical problems. For example, as there is a greater concentration of salts in sea-water than there is in animal tissues, they tend to invade the animals' bodies in an attempt to establish some sort of equilibrium. The animals' survival depends on their ability to pump the excess salts from their bodies. This entails the expenditure of energy—the business of the mitochondrion.

There are different ways of getting rid of waste material. It can be excreted via the kidney and out with urine or its equivalent, or it can be deposited as a kind of shell on the outside, as with the crab, or as crystals within the skin. The skin has a very large surface area and if inert crystals can be laid down in it they will also be effectively removed from the system.

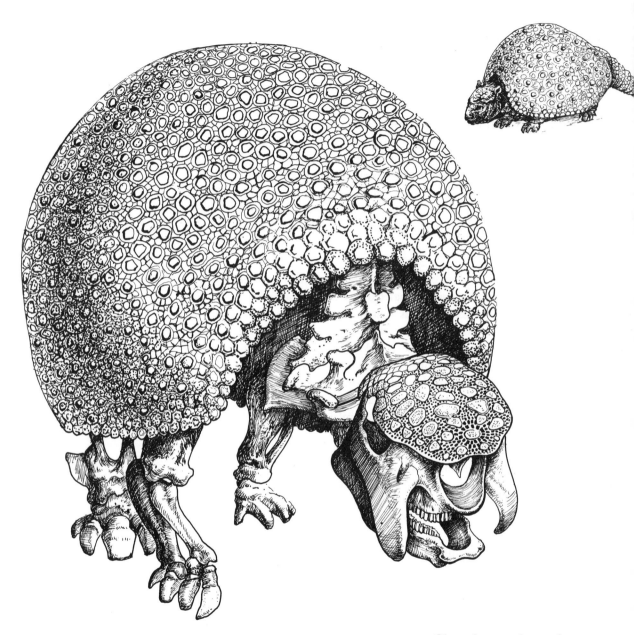

Glyptodon—*a giant extinct armadillo from South America, the size of a small car.*

From being at first merely a waste product, these mineral deposits in the skin must have rapidly acquired a new role. Phosphates are essential for life, yet they are in short supply in the sea, in which there is a phosphate cycle; a seasonal abundance and dearth. The acquisition of a phosphate store would be of immediate survival advantage to an organism. During the period of phosphate availability, the first vertebrates would have excreted the excess calcium in the form of apatite, as the most convenient way of neutralising the effects of calcium. The mitochondria could hardly have avoided doing this, and the consequence was that these animals acquired, almost by accident, a phosphate store which would have served them in good stead during the seasonal dearth.

Natural selection would have ensured that the individuals with the bigger stores stood the better chance of survival. This led to the development of a proper armour which would have served as a protective covering—yet another important advantage to creatures feeding on organic debris in mud and at the mercy of any predator that happened along.

The protective function of bone is one that has remained with all the vertebrates—our brains are particularly well protected by our cranium. Indeed, many animals that have taken to a rather inoffensive way of life, such as armadilloes, have a real need for protection. Such creatures have re-evolved a bony carapace. Perhaps the most dramatic example is the giant South American armadillo, *Glyptodon*, which only became extinct in comparatively recent times. *Glyptodon* was almost the size of a small car. When the specimen was being sketched in the Natural History Museum in London, it seemed to be the favourite of the children. One little girl, who vouchsafed her affection for it, confessed that she liked his ears (the bony flanges on the skull which give it a rather sad doggy appearance). The reaction of two little boys was simply that it was like school mashed potatoes—with all the lumps!

But to return to thé ostracoderms: they lived in water so there was no need for them to have the support of a bony internal skeleton—a cartilaginous skeleton was perfectly adequate. Incidentally, sharks do not have true bones but cartilage impregnated with calcium phosphate. This is termed calcified cartilage and is quite different from true bone. Nevertheless, in certain groups of fish, the bone developing in the skin began to invade and replace the cartilage of the internal skeleton so that gradually a bony internal skeleton evolved. This had very important consequences later on when these fish ventured onto land and were capable of supporting themselves in air by virtue of their bony internal skeletons. It was the possession of such a skeleton that enabled the vertebrates to

13

conquer the land and achieve the evolutionary stage at which we now find ourselves.

A kind of potted version of this evolutionary development can be seen during our development in the womb. At first the foetus has an internal skeleton of cartilage, certain areas calcify and the cartilage begins to break down during the invasion of bone-forming cells, or osteoblasts, and as the cartilage disintegrates so true bone replaces it. The cartilaginous skeleton of an embryo does not gradually *change into* bone but is *replaced by* bone. This observation was made in the 1700's by the famous anatomist John Hunter.

Once the vertebrates were established on land, bone played the role for which it is now best known, that is of an internal support, allowing the limbs to move and the body to be held off the ground.

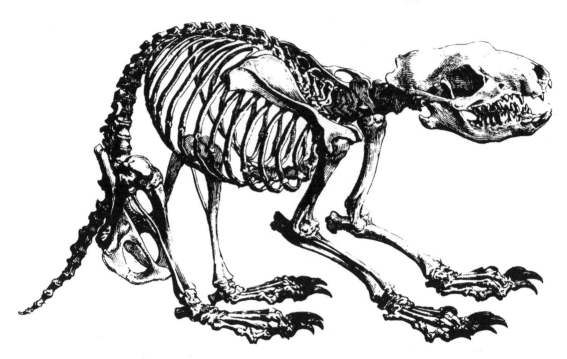

Skeleton of a hedgehog from William Cheselden's Osteographia, *1733.*

But this is not the whole story, because, as we mentioned earlier, the skeleton now fulfils yet another role as a chemical store for calcium. Initially bone was formed through the animals' need to get rid of calcium; it now forms an important store of calcium. Because we no longer live in the sea, we have to seek out calcium, one of the best sources of which is milk.

Of our skeleton's several functions—an internal support, a protection, and also a chemical store—it is possible to argue which role is the most 'important'. In evolutionary terms some sort of chemical store came very early in the history of bone, followed by protection, then by support. In old age this system appears to go into reverse. Commonly in old people, the bones become rarified, spongy and brittle, frequently allowing fractures, particularly of the neck of the femur at the hip joint, to occur. This is the particular fracture from which Sir Winston Churchill suffered late in his life.

A similar ageing process seems to occur when people are in a state of weightlessness over an appreciable period, a very serious problem for astronauts. There is at present considerable consternation among American and Russian space officials over this particular medical hazard. Both have discovered that during space voyages the astronauts begin to lose their skeleton, the bone being slightly, but nonetheless appreciably, reduced and calcium salts are excreted. This is not too serious a problem for very short space journeys, but on long space journeys to other planets in the solar system, and not just our nearest satellite, there are very grave difficulties to be overcome. It is perfectly feasible for long space journeys to be made, but the astronauts would not be able to return to earth because the deceleration involved in re-entering the earth's atmosphere would be such that the entire skeleton would just shatter. It is almost as if the body is once more back in the conditions experienced by the earliest marine vertebrates at the very beginning of our evolutionary history, i.e. in a state of relative weightlessness. It may therefore be necessary to provide future 'long-distance' astronauts with an artificial gravitational field in order that such regressive bone changes do not occur in flight.

3 Bones in action

Bones become pliable when the mineral matter is removed.

Bone makes an ideal internal supporting material by virtue of its amazing strength. It is a two-phase material, comprising two contrasting substances: the fibrous protein collagen (which is strong in tension) and the mineral apatite (which is strong in compression). The crystallites of calcium phosphate are exceedingly small and are aligned along the collagen fibrils. The way in which these two substances are actually bound together is still not properly understood. If the mineral matter is removed by acids, the result is a rubbery bone that is so flexible it can be tied in a knot. If the organic matter is destroyed a brittle bone results. In some way the mineral matter locks in the protein, so that the organic matter is able to survive for millions of years. Presumably the close packing of the crystallites seals off the organic matter from natural destructive agents.

The strength of bone has been the subject of much study, since on theoretical grounds it ought not to have the strength it undoubtedly does have. It is frequently compared to reinforced concrete but this analogy is not very helpful. The real problem is that a material composed mainly of crystalline matter is expected to be exceedingly weak under tension. Most stresses on the skeleton are likely to be compressive but there are sufficient tensional forces operating to ensure failure. All minerals have faults in them and no crystal is perfect, for in the packing of molecules there occur slight dislocations. If one imagines a pack of cards, it is easy to envisage part of the pack sliding over the other. On the ultramicroscopic scale this is what happens in crystals. If the material is subjected to compressive forces it will not matter, for if a tiny crack develops, these forces will close it up. In contrast, if tensional forces are applied, the crystal is pulled in opposite directions, any molecular dislocations will become exaggerated and the crack will not only open up but will spread through the material, which will thus fail. The formation of these 'Griffith cracks' (so-called after their discoverer) are of very great importance in all structural materials, and many people have lost their lives because of them. During the early days of the aircraft industry, it was discovered that, for no apparent reason, flying aeroplanes would suddenly fall apart; the materials of which they were constructed would quite inexplicably collapse. Since this always happened while the planes were in flight, it naturally caused much consternation. These mysterious happenings were finally traced to the gradual spread of 'Griffith cracks' through the metals used in construction.

This problem has been overcome in a novel way in fibre-glass. The brittle material, in the form of fine glass fibres, is embedded in epoxy resin. The fibres themselves are probably too small for 'Griffith cracks' to form, but even if they were to do so, they could not possibly spread through the entire structure. As a crack develops in a fibre, it will spread into the resin which deforms to take up the strain. Dr John Currey believes that this modern structural material provides the closest parallel to bone, in which there are small, brittle crystals of apatite set in a matrix of deformable collagen. The parallel is certainly there, but it seems to us that the packing of the crystallites is not the same as in fibre-glass.

The key to the strength of bone lies, as John Currey recognised, in the mineral component being in minute crystallites. If there is a mass of minute crystals packed together, whenever a crack develops in one crystal it will not spread to its neighbours—the force merely passes around the margins of the adjacent crystals and is thus absorbed. The Cook-Gordon crack-stopper effect also comes into play; as a crack develops the force generated will open up a further crack at the interface of an adjacent crystal and at right angles to the original crack. This will effectively prevent any further spread, since it is almost impossible to extend a crack by applying a force at right angles to it. This is the reason why bone is so strong. A fortune probably awaits the person who manages to develop a structural material based on the same plan as bone. So far it has not been done but undoubtedly someone, sometime, will accomplish it.

Although bone is ideally suited for a structural material, like any building material it has to be put together in the proper manner. To be effective, its internal architecture must be appropriate for the function it has to perform. For support of a weight by a given volume of material, it is more efficient that it should be organised in the form of a cylinder rather than as a solid chunk. This is normal engineering practice and this is how the long bones of the limbs of animals are constructed.

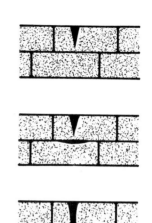

How a crack is stopped from spreading.

Section of a bone showing spongy middle and solid edge, used by Eskimos to cut blocks of snow for their igloos.

The solid compact part of limb bones is the outer cortex of the mid-parts of the shafts. At the ends of the bones, where the bones approach the joints, the situation is entirely different and the bones are of a loose, spongy texture. The ends of even the most massive bones can be easily gnawed by most animals with a taste for such activity. When examined closely this spongy looking tissue is, in fact, well organised, showing a delicate internal architecture of which an ecclesiastical Gothic builder would not be ashamed. There are series of narrow beams, or trabeculae, joined by minor cross struts which together form an intricate three-dimensional scaffolding. The main trabeculae line up along the major axes of force to which the bone is subjected during development. When the weight of the body is transmitted from one bone to another, as at the knee or hip joint, for example, the forces are distributed through this tracery of trabeculae to the compact cortex of the outer parts of the shafts. This complexity at the ends of the bones is necessary simply because both movement and weight have to be conveyed with the bones in different positions. In whatever manner the limbs are displayed, within reason, the forces can still be effectively distributed through the trabeculae.

At the hip joint, the articular head of the femur fits into the socket laterally, being set almost at right angles to the main shaft of the bone. The trabeculae carry the forces from the rounded head to the neck of the bone and thereafter to the compact cortex of the shaft. The entire weight of the body must be supported through the delicate bony struts of the neck of the femur. It is at this particular place that old people all too frequently fracture their limbs.

In birds, the light hollow wing bones are strengthened by fine cross-wise struts; the internal supports in aeroplane wings, in this respect, are identical. In all the examples one may care to choose, it is evident that the internal architecture of bones is based on efficient engineering principles. Or perhaps one should say that efficient engineering principles are merely copies of what nature has already achieved.

The overall shape of a bone and its internal architecture are a reflection of the forces acting upon it. By far the most dramatic illustration of this is the bipedal goat described by Professor E. J. Slijper. This little animal was born without any front legs and hence had to learn to walk about on two. This it did. But the striking feature of its backbone was that it was quite unlike that of any other goat. The arrangement of the spines of the vertebrae was quite different, and the overall vertebral column developed a sinuous curvature, somewhat reminiscent of a human's—another bipedal animal. This goat is an extreme example, but during the life

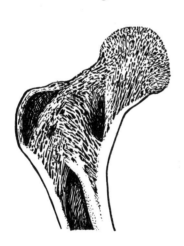

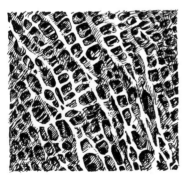

The head of a human thigh bone cut to reveal the fine trabeculae, which distribute the forces applied to the bone. Below: close-up showing delicate connecting struts.

of any individual the forces acting on the skeleton do not remain constant. If the animal is to function properly, it must be capable of altering the shape of its bones, albeit slightly, to accommodate the changing requirements. And this does indeed happen. This is obvious if a limb is broken, even more so if one suffers a greenstick fracture where bone splinters into many slivers. If correctly set, the limb will heal and eventually the new bone will be as good as new.

During the 1914-1918 war, many soldiers were wounded in the face by bullets and shrapnel. Their jaws were often shattered into tiny fragments. Warwick James performed miracles on these men; he wired up the shattered pieces of bone and eventually the jaws regrew, so that after 20 years there was no evidence that at one time the jaws had been smashed to smithereens.

These examples emphasise that bone is a living tissue capable of regeneration. For a bone to change its shape in response to changing conditions, as well as being capable of producing new bone, it must at the same time be able to destroy bone. Whereas there are special cells which produce bone, the osteoblasts, there are also cells, the osteoclasts, whose function is the destruction of bone.

An example of what these two types of cell can accomplish when working in concert has been demonstrated by Professor J. Trueta. Some years ago a patient was referred to him with a diseased thigh —the head of the femur was dead and the advice of the pathologists was that it should be excised. From his knowledge of the behaviour of bone, Trueta believed that the cells of the body could deal with the problem, if given suitable encouragement. He inserted into the dead bone a sliver of healthy bone, removed from another part of the femur. The live bone grew and as it did so, osteoclasts removed the dead bone, so that gradually the dead bone was completely replaced by healthy. Today the patient cannot remember which hip was diseased.

In the normal course of events, the remodelling that goes on in our bones is imperceptible. Bone is constantly being resorbed and renewed. There is a delicate equilibrium maintained, which only begins to break down in old age, when the amount of resorption is greater than redeposition. The bones become more porous, resulting in a condition known medically as osteoporosis. They are then less able to withstand the strains normally placed on them, they become brittle and also take much longer to heal, should they get broken.

Where two bones move against each other in a joint, as at the knee or elbow, the joint surface needs to be properly lubricated. In fact, all the basic engineering principles are applicable here. First of all the joint is enclosed in a capsular ligament, and in the small space between it and the bone is synovial fluid—the lubricant. The articular surfaces of the bones are covered with cartilage which acts as a bearing surface. As with oil, the fluid is incompressible and movement is virtually frictionless, a property which is absolutely vital as otherwise heat will be generated and the bearing surfaces destroyed and thereafter the actual bones. In the joint, the lubricant is forced between the two surfaces and any heat generated is carried away. The thinner the layer of lubricant the faster the movement possible, but if it is too thin, the bearing surfaces will come into contact, with unfortunate results.

The nature of the bearing surface material is critical: it must be smooth, must allow the lubricant to adhere to it and, finally, must have shear-strength. Identical problems, and solutions, are found in the big end of a car where the crank shaft 'articulates' with the connecting rod. Lubricating oil is pumped from the sump through fine oil-ways. The crank shaft is covered with a bearing material of white metal or 'babbitt' named after the man who invented the alloy in 1839. (Typically it comprises 80% tin, 8% antimony and 4% copper. However, these alloys may range from 90% tin and no lead, to 80% lead and only 5% tin). The oil adheres to these alloys. The bearing surface of the connecting or 'con' rod is composed of similar metal bearings but in the form of half shells to facilitate replacement. If a driver 'runs his big end', i.e. drives his vehicle with insufficient oil, the heat generated tears the surface of the bearing-metal shell. When this happens in a human joint and the cartilage-bearing surface is plucked off, it is termed osteoarthritis. It is pleasing to think of cars also suffering from an equivalent ailment!

So far we have discussed the nature of bone itself and its basic structure. We have mentioned the forces to which bones are subjected, but these will vary enormously depending on the particular animal studied. One of the most obvious facts about animals is that they are of different sizes. The size of an animal will reflect to a very large extent the nature of its internal skeleton and the overall shape of the animal. J. B. S. Haldane in his famous essay 'On being the right size' discussed the extra demands on the skeleton involved in increases of size.

In John Bunyan's *Pilgrim's Progress*, Christian met with Giant Pope and Giant Pagan. They were ten times higher than he was, and ten times as thick. Their weight was 1000 times his but the cross-section of the limb bones only a hundred times. As Haldane wrote: 'every square inch of giant bones had to support ten times the weight borne by a square inch of human bone. As the human thigh-bone breaks under ten times the human weight, Pope and

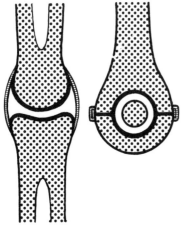

Joint with bearing surfaces and lubricant, together with crank shaft and connecting rod of a car's big end to show basic similarity.

Pagan would have broken their thighs every time they took a step. This is doubtless why they were sitting down in the picture I remember. But it lessens one's respect for Christian and Jack the Giant Killer'.

If a small graceful creature like the spindly-legged gazelle increases in size, it would have to develop thick legs like an elephant's, because every pound of weight must be supported by the same area of bone.

These problems are the simple consequences of the fact that the linear increase in size will result in the surface area increasing by the square and the volume (and the weight of the animal) by the cube. Haldane illustrated the effect of gravity on size as follows: 'You can drop a mouse down a thousand-yard mine shaft; and, on arriving at the bottom, it gets a slight shock and walks away. A rat would probably be killed, though it can fall safely from the eleventh storey of a building; a man is killed, a horse splashes'.

Apart from size, the form of the skeleton will also be determined by the animal's life-style. The different proportions of the bones of the limbs tell us whether the creature is a fast runner or a powerful digger of holes. The movement at the main joints of the limbs are of the type termed third-order levers. The fulcrum is at one end, the weight at the other and the force somewhere in between. A good example of this is when something in the palm of the hand is lifted by raising the forearm at the elbow. The elbow joint is the

Human elbow joint as a second order lever. Badger's forelimb specialised for powerful digging and cheetah's for fast running.

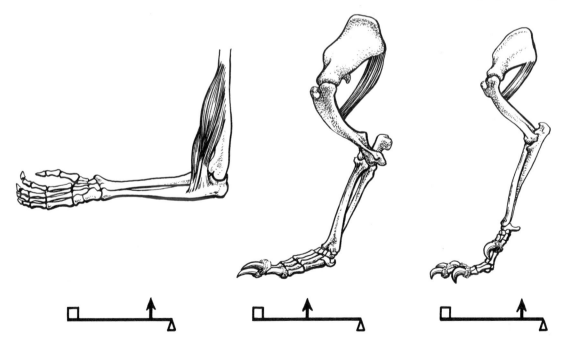

fulcrum and the biceps muscle provides the force which is exerted to lift the hand. The further away from the fulcrum that the force is applied, the easier it is to move the weight. This situation constitutes a good 'mechanical advantage'. A familiar example of this is opening or closing a door; it is easier to move a heavy door if one pushes it at the point furthest away from the hinge, but even a light door is difficult to open or close by applying pressure close to the hinge. The moving of a heavy door will be fairly slow but a large weight will have been shifted. Limbs built to this specification will have long humeri (upper arm bones) and femora (thigh bones) and the muscles from the shoulder and hip joints connected to these bones will be similarly long. Such limbs will be comparatively slow in action but extremely powerful. Badgers, anteaters and aardvarks (the African earth pig) have limbs of this sort—all are powerful diggers, none are notable for their turn of speed.

At the other end of the scale a moving joint or limb can have a poor mechanical advantage. To return to the analogy of the door, if it is sufficiently light and is pushed only a short distance from the hinge the edge furthest away will describe a very large arc and will do so at a speed many times that at the point of application of the force. The swishing of a cane is perhaps a better example. By small movements of the wrist a cane can be made to describe a large arc with great speed. A limb based on this scheme would be thin and the main muscles from shoulder and hip would be short. Animals with these kinds of legs would be specialised for speed. Examples of such are the cheetah (the fastest land animal of all), horse, deer and antelopes. These animals can run fast but their legs are not powerful. These terms are of course comparative, and it would be difficult to convince anyone who has been kicked by a horse that

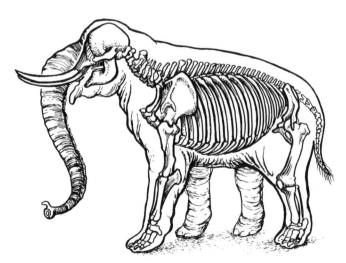

Elephant with pillar-like limbs to support its own weight.

the beast was weak in the leg. It seems more powerful than it really is by virtue of the speed of delivery.

With long, and often spindly legs, the stride is lengthened; the arc described by the limb from movement at the shoulder or hip ensures speed. The stride is further improved in the cheetah by flexure of the backbone; this gives the animal an extra 6 m.p.h. Some zoological wag has claimed that an amputated cheetah could 'run' at 6 m.p.h. by means of flexing its backbone alone.

The backbone of the Hero shrew, *Scutisorex*, is specialised for weight bearing: not its own, but rather someone else's. It is claimed that it can bear the weight of a twelve stone man and survive the ordeal fairly happily. These tiny animals spend their time burrowing under rocks and boulders and have developed a most peculiar looking backbone, with numerous nobbly growths of bone on the sides of all the vertebrae. This specialisation presumably prevents them being crushed by collapsing boulders.

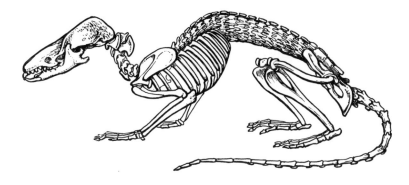

Hero shrew with knobbly backbone to support the weight of others.

A quite different specialisation that has occurred in a number of disparate groups is bipedalism, and by far the most familiar example of this trait is man. Because our ancestors lived in the trees and moved around, Tarzan-like, by means of their arms they would have arrived feet first on landing. When the forests were deserted by our distant ancestors for the more open savannah country, walking upright would have had the advantage of enabling them to survey the landscape for possible predators or prey. The main disadvantage of this gait is instability. A four-footed animal does not easily lose its balance—with a two-footed one it is all too easy. Fortunately, survival in a tree habitat puts a high premium on the sense of balance, and the degree of muscular coordination required in negotiating branches would have also stood our ancestors in good stead. They were in fact *preadapted* for bipedalism. The part of the brain concerned with muscular coordination and balance— the cerebellum—is highly developed in man.

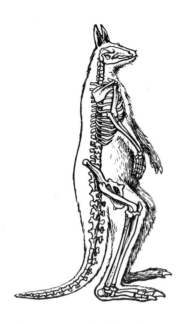

The kangaroo has solved the problems of being a biped in a quite different way. It leaps about with both feet in unison while its strong muscular tail acts as a counterbalance to the rest of the body. When in a resting position, but not lying down, it can squat back on its tail. This is the position it takes up when it is persuaded to enter the boxing-ring, where it seems to be able to acquit itself with honour.

Birds too can be considered as bipeds since, when grounded, they run about on their hind legs. Unfortunately, most birds do not have a long bony tail that is designed to act as a counterbalance. They did originally, the first bird *Archaeopteryx* having had such a tail; but as its descendants evolved, they lost it. The tail may have acted as an automatic stabiliser to correct pitch when gliding, but its loss in the process of reducing weight, also led to the birds being unstable in the air. This is, in fact, an advantage since it gives the animal a lower stalling speed and makes them more manoeuvrable. Of course this only works if the eyes and organs of balance, together with muscular coordination, are improved. For a bird to progress overland, the legs have to be brought forwards under the

Kangaroo—a biped resting as a tripod.

The first bird, Archaeopteryx, with its long bony reptilian tail, and a present-day rook.

centre of gravity; and to achieve this the femur projects forwards almost horizontally. The leg in fact moves from the knee joint, which is centred just below the centre of gravity. As a bird walks, and each leg alternately takes the entire weight of the animal, it has to swing inwards to maintain balance. This particular gait is best seen in ducks and geese, with their characteristic waddle.

The extinct carnivorous dinosaur *Tyrannosaurus rex*, like the kangaroo, had a large muscular tail which acted as a counterbalance. However, *Tyrannosaurus* was a very large animal indeed, standing 15 feet high and any possibility of leaping about like a kangaroo was quite out of the question. One small kangaroo-like dinosaur is actually known, from South Africa, but evidence for its existence comes only from fossil footprints. *Tyrannosaurus* was obliged to plod along one foot after the other, and the majestic striding gait portrayed in such films as *One Million Years B.C.* never happened. We know from its footprints that by taking very short pigeon-toed, mincing steps, *Tyrannosaurus* waddled. The femur projected forwards somewhat, and movement of the limb at the hip joint was only through a very small arc. Like birds, the main stride—not that there was much of it—was from the knee joint.

The giant carnivorous dinosaur, Tyrannosaurus rex, *on the move and lying down.*

Tyrannosaurus is always portrayed standing up, but it is inconceivable that it spent its entire life erect. There must have been times, particularly after having gorged itself off a rotting corpse, when it simply had to lie down. Not until very recently has it been portrayed lying down, nor had it ever been considered how it got up again. Mr Barney Newman has pondered these two problems, and though his new restoration is not so sensational in appearance, from a biological point of view it appears much more realistic. He has convincingly demonstrated that the ridiculous, supposedly vestigial, fore-limbs were not just the last remnants of vanishing front legs, but they also served a very important function in allowing the animal on the one hand to pick bits of rotting meat out of its teeth, and on the other to get up. If one considers a prone *Tyrannosaurus*, whenever it stretched its legs, it would only succeed in pushing its belly along the ground. The fore-limbs had a massive musculature and could dig into the ground and hold firm. In this situation, when the hind-limbs were extended, the two-pronged fore-legs would prevent any forward sliding of the body. In this way, the force would be transmitted in such a way that it would have lifted the body from the ground. A backward toss of the head would also have aided the process of getting up.

One of the very earliest prehistoric animals (or rather a fragment of one bone) to have been described and drawn was a close relative of the carnivorous *Tyrannosaurus*. This specimen was described by Dr Robert Plot in his *Natural History of Oxfordshire* in 1677, and he pointed out that it was the kneeward end of the femur. He knew it was not part of an elephant, and for some strange reason suggested it might have belonged to a giant man. (It was believed at that time that there had been giants of men). Nearly a hundred years later Dr R. Brookes copied many of Plot's figures and more or less repeated many of his descriptions. But what he did in his book was to label this dinosaur bone '*Scrotum humanum*', a name recalling the bone's general appearance. At the end of the femur, there are two condyles (rounded articulating projections), which gave the bone the appearance of a gigantic petrified scrotum, presumably from a victim of elephantiasis. The bone in question was two feet in circumference. The odd feature was that this name *Scrotum humanum* was introduced after the official start of zoological nomenclature in 1758, and therefore it is the first valid name ever given to a dinosaur. Unfortunately the International Commission of Zoological Nomenclature has ruled that if an animal, be it living or fossil, has not had the original name used in the last fifty years then that name must be dropped. We have the ironic situation that the name *Scrotum humanum* is a *nomen oblitum*, a forgotten name.

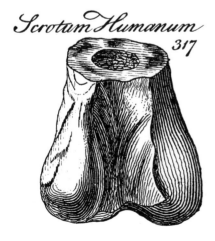

The end of a dinosaur's thigh bone, first described by Robert Plot in 1677 as coming from the leg of a giant man.

The other kinds of dinosaur were vegetarian, and some of the more successful were also bipedal. These are termed facultative bipeds because they could walk about on either two legs or four, just as the mood took them. As well as looking exceedingly impressive, the dinosaurs present us with many unsolved problems, not the least of which is why they died out. There are dozens of theories about this, but the plain fact is nobody actually knows. Of one thing, however, we can be perfectly certain and that is that mankind was entirely innocent in the matter. He did not make his appearance on earth until about 65 million years after the last dinosaur had disappeared.

At the time of writing a new controversy has arisen regarding the enormous *Brontosaurus* and its allies. These largest of land animals are generally thought to have spent most of their time in swamps feeding on soft vegetation and at the same time taking the weight off their feet. An American, Dr Bakkar, has now offered the suggestion that they were true land dwellers. But, the controversy

Herbivorous dinosaurs from 170 million years ago.

that has arisen over this seems rather pointless, for we have known for a very long time that these animals could walk about quite happily on land, and they certainly laid their eggs on land. Furthermore, we have fairly good evidence that they did spend a lot of time in water; suites of footprint trackways have been found, and in some cases these consist of the prints of the front legs only. In other instances only the back legs have left prints and furthermore, the long sinuous tail left no trackway at all. This situation is quite incomprehensible on land. No one would deign to suggest that huge animals like *Brontosaurus* pranced about on their front legs and never let their tail dangle in the dust. These trackways only make sense if the animals were immersed. Buoyed up by the water with their tails floating, they could trot along just on their front legs, probing perhaps into the mud or sand, or at least dipping into it; alternatively they could rely on their hind legs. This sort of interpretation easily explains the trackways which have been found. There is no doubt that these giant herbivorous dinosaurs could go both on to land and into the swamps, and the proportion of time they spent in one or the other is something we will probably never know.

As well as dinosaur bones enabling us to reconstruct their skeletons, impressions of mummified skin telling us what they looked like on the outside, and fossil footprints indicating how they got around, we can study the bones under the microscope. In a number of dinosaurs the bones show annual growth rings and it becomes possible to answer the question 'how many years did a dinosaur live?' We know they hatched from fairly small eggs because these have been found as fossils; we also know they reached gigantic sizes. Professor Glenn Jepsen has counted 120 growth rings in some specimens so we can be certain that some dinosaurs lived for at least 120 years. In view of the longevity of such reptiles as the giant tortoises, a hundred years or more seems to be a pretty reasonable estimate.

An ichthyosaur—the prehistoric fish-lizard.

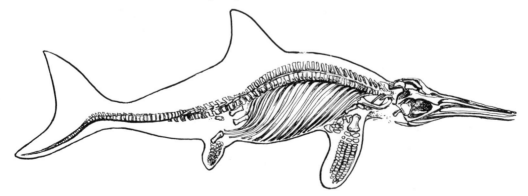

During the age of dinosaurs there were also strange giant reptiles inhabiting the seas. These were the ichthyosaurs, which looked rather like dolphins with large dorsal and tail fins. They never left the sea and gave birth to live young, unlike normal reptiles which lay eggs on land.

Another group of marine reptiles were the nothosaurs. These had webbed feet and a long dorsal fin on the tail, and evolved into the plesiosaurs which at the beginning of the last century were aptly likened to a serpent threaded through a turtle. These creatures had a barrel-shaped body, four limb-paddles, a long neck, a small head and a short tail with a dorsal fin. Professor D. M. S. Watson had determined their swimming behaviour from a detailed study of their skeletons. Apparently, these animals spent most of their lives in the surface waters feeding on fish. They were incapable of diving but were exceedingly adept at manoeuvring—something akin to dodgem cars. They could make extremely rapid flicks with their paddles, either in normal swimming strokes or in backing strokes, and because of this they could twist and turn very rapidly, even though they were not powerful swimmers.

Every few years the notion is put forward that Scotland's Loch Ness is inhabited by plesiosaurs and some of the descriptions of 'Nessie' bear a faint resemblance to a plesiosaur—if one looks at them with the eye of faith. Unfortunately, the described behaviour of the Loch Ness Monster is quite incompatible with its being a plesiosaur. A recent colour cine-film of the Monster, taken by a Professor of Chemistry, shows it very close to shore breaking the surface in two humps and then going straight down again. A friend of ours has been to the bottom of the loch in a submarine and has seen these objects rising to the surface. They are gas-filled vegetation mats, as Dr Maurice Burton postulated over ten years ago.

There is another type of plesiosaur known, the pliosaur, which has a short neck and large head. They hunted squids and cuttle-fish, were powerful swimmers and could dive after their prey. Again, their swimming movements have been deduced from a detailed study of the skeletons. The main swimming stroke was by driving the paddles downwards at an angle so that they acted as hydrofoils. Barney Newman, from a study of the hydrodynamics of their swimming, has concluded that the main stroke would drive the animal forwards and upwards, thereafter the body, while still moving forward, would sink and the form of the paddle would cause it to rise automatically until it was positioned for the next power stroke. In this way the pliosaur would travel in an undulating path in exactly the same way as Dr Peter Purves has demonstrated for the living dolphins and whales. All of them 'porpoised'.

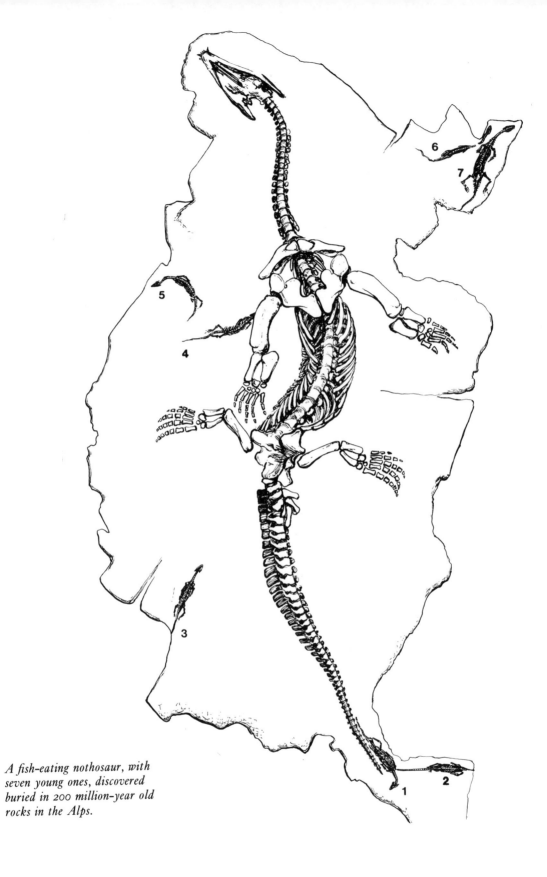

A fish-eating nothosaur, with seven young ones, discovered buried in 200 million-year old rocks in the Alps.

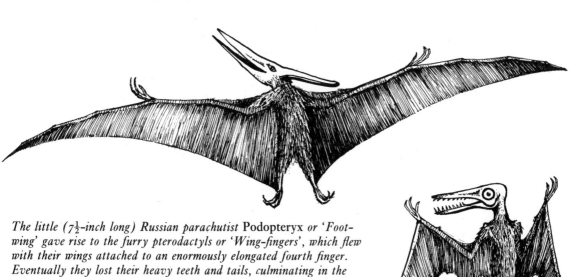

The little (7½-inch long) Russian parachutist Podopteryx *or 'Foot-wing' gave rise to the furry pterodactyls or 'Wing-fingers', which flew with their wings attached to an enormously elongated fourth finger. Eventually they lost their heavy teeth and tails, culminating in the gliding* Pteranodon *with its 27ft. wing span.*

As well as dominating life on land and sea, the reptiles also con-quered the air. Long before birds had evolved the skies were populated by strange leathery-winged creatures, which are usually described as flying *reptiles* (pterosaurs). It is perhaps a mistake to think of them as reptiles, which are cold-blooded, scaly animals, because the pterosaurs were undoubtedly warm-blooded and Dr A. G. Sharov of Moscow has in his collection a pterosaur *Sordes pilosus* or 'Filthy Fur' on which a covering of thick fur is perfectly preserved. The pterosaurs had air-filled bones as do birds, their brains were birdlike, and the only thing they did not have were feathers. The wings were made of a membrane stretched along a greatly elongated fourth finger and fixed to the back legs.

Dr Sharov has even more recently described the ancestor of these strange creatures. Contrary to all expectations it has a complete gliding membrane between its back legs and tail with nothing much to speak of at the front end. Nonetheless from such be-ginnings the furry pterosaurs evolved. They gave rise to the largest flying animal the world has seen, *Pteranodon*, with a wing-span of 27 feet, but the wing bones of this colossus were only as thick as a post card!

How this giant managed to land and take off has always been something of a mystery. Recently Cherrie Bramwell has put data from fossil skeletons through the computer and has been able to determine the landing speed at 12 m.p.h., the optimum flying speed 15 m.p.h. and the maximum speed 35 m.p.h. It has emerged that *Pteranodon* was able to fly at very low speeds and that at a speed of 15 m.p.h., it had a low sink of only 1 m.p.h. and was highly manoeuvrable in consequence. The fastest speeds were

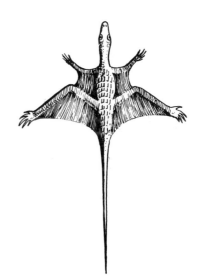

achieved with the wings swept back—the fore-runner of the swing-wing plane.

It may be wondered why the pterosaurs should have been ousted by the birds, in view of the excellence of their adaptions for flight. The answer lies in their problems on land. The wings would have been partly folded over their backs and they would have walked with difficulty on all fours. In contrast, when birds alight, the wings are tucked out of the way and they can run about as terrestrial bipeds. Indeed, after the extinction of the dinosaurs, some giant secondary flightless birds became, for a time, the major carnivores on earth.

The detailed picture of life during the age of dinosaurs has been acquired almost entirely as a result of the study of fossilised bones. Skeletons and scattered isolated bones have a fascinating story to reveal of past life on our planet. We do not have a time machine but at least fossil bones allow us to enter into some of the mysteries of earth history.

Odd bones 4

The bones of the skeleton are primarily concerned with support and locomotion. The limbs, whether of bipeds or quadrupeds, are the means of their getting around. But legless animals also seem to manage fairly well, as anyone who has tried to catch a snake can testify. Snakes have a most unfortunate reputation, presumably because some of them poison and inject digestive juices into their prey. The digestive enzymes attack the prey's blood and nervous system having the effect of immobilising the prey and poisoning it to death.

A creature as serpentine as a snake must have some mechanism preventing its body from twisting on itself. If a snake simply

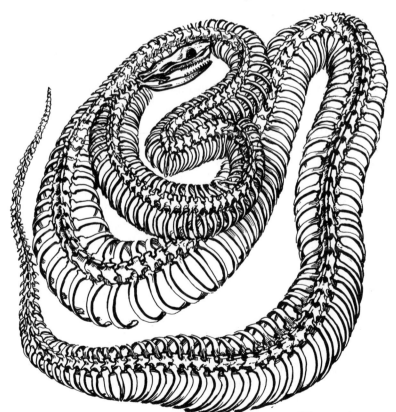

Skeleton of a python.

twisted like a rope, it would cause havoc with its blood vessels and nerves. To prevent such a catastrophe the vertebrae, of which there are between 200 and 400, are specially constructed. All vertebrae of land animals articulate with one another by means of two pairs of projections, the front ones face upwards and fit onto the back ones of the vertebra in front, which face downwards. By means of these joints (zygapophyses) a degree of movement becomes possible. These vertebral joints allow us to bend and twist our backs. In the snake the backbone is restricted to a sideways motion, allowing it to wriggle along the ground.

A snake held in the hand can project its body horizontally in the air, apparently defying gravity. There is plenty of lateral, but virtually no vertical, movement as it surveys the scene. This phenomenon is due to the presence of an extra set of articulating facets which restrict any up-and-down flexing to 30°, but allow side-to-side movements of twice this amount. On the front surface above the zygopophyses, these extra facets (zygosphenes) face downwards. The posterior ones (zygantra) face upwards and articulate with the anterior facets. In egg-eating snakes, there are also projections on the underside of the neck vertebrae which break eggs as they pass down the oesophagus. The snake's movements are restricted to the horizontal plane by the nature of the modified vertebrae. The weight of the animal is supported by the rest of the skeleton which consists of long ribs. Some snakes are even capable of crawling along, albeit incredibly slowly, by means of the ribs alone.

Although the vertebral column and ribs are modified for a highly specialised mode of life, their function is nevertheless comparable to the skeletons discussed in the previous chapter. There are other bones in animals which fulfil quite different roles. Among these odd bones, those of the vertebrate ear have by far the most unusual history.

In the middle ear there is a set of three tiny bones: the hammer, anvil and stirrup; or in scientific terms malleus, incus and stapes. The ear cavity is connected to the throat by a fine tube—the eustachian tube. When one holds one's nose and blows whenever experiencing discomfort in the ears, during air travel for example, one is blowing air into the middle ear cavity; this equalises the pressure on both sides of the ear drum.

The ear drum vibrates when sound waves reach it; this in turn vibrates the malleus and the other two bones. As the vibration travels from bone to bone, it is amplified and finally transmitted to the inner ear, where it is picked up in the fluid-filled cochlea. The bones in the ear therefore constitute a sound transmitting and amplification system.

The wriggling of a snake is restricted to a sideways motion by extra articulations on the vertebrae.

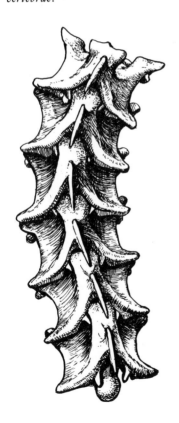

In our far distant ancestors these bones were originally part of a breathing apparatus, then they became concerned with feeding, and only recently with hearing. In the earliest vertebrates (ostracoderms), the gills, by which the animals respired, were supported by skeletal structures, the branchial arches. Each of these arches was made up of a series of bony elements and were hinged halfway down, allowing some movement of the gills. These ostracoderms did not have any jaws but their descendants acquired bony jaws, the first gill bar evolving to form a hinged mandible. The two bones that formed the jaw joint were the quadrate in the upper jaw and the articular in the lower. The expansion of this mandibular arch resulted in the gill immediately behind it becoming compressed. The upper part was transformed into a 'pocket', the spiracle. (This is the opening behind the eye in sharks and rays.) All that was left of the lower part of this gill was a groove in the lower part of the mouth. This still survives in man as the housing of the salivary glands.

With the elimination of the first gill, the uppermost bone which had supported it articulated with the skull and jaws. In the primitive jawed fishes the entire jaw was suspended from the skull by means of this bone—the hyomandibular.

When fish evolved into the amphibians, and later the land-dwelling reptiles, further changes occurred. The upper jaw became fused with the cranium and the hyomandibular, although it was still attached to the upper jaw bone (the quadrate) and no longer played any role in suspending the jaws. The spiracle was eventually replaced by a notch in the skull across which a membrane was stretched. This was linked to the skull by the hyomandibular bone, which at this point palaeontologists refer to as the stapes. In the reptiles this same bone is often called the columella. Here at last we have a bone which can conduct sound waves.

This arrangement for picking up sounds is common to all land-living vertebrates. But as we have already mentioned, in ourselves and all other warm-blooded furry creatures, there are *three* bones in the middle ear. From the study of fossils and embryos, we know that the extra two bones were once part of the jaw joint of our distant ancestors. The question has been asked for a long time: how did the jaw joint become incorporated into the ear? Indeed, this change seems so fantastic that it has sometimes been used as one of the major planks in the anti-evolutionists' case against

Ostracoderm

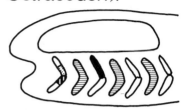

Fish

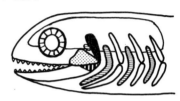

Amphibian

Mammal-like Reptile

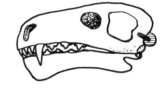

Mammal

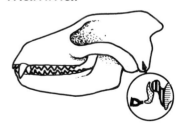

The history of ear bones. In the ostracoderm they supported the gills, in fish one bone suspended the jaws from the skull, the other two formed the jaw joint. In amphibians and the reptiles there was a single sound-conducting bone in the ear. Finally in the mammals the two bones of the jaw joint linked up to form the string of three ear ossicles.

Darwinian evolution theory. If this really happened, it is asked, 'how did the creature contrive to feed *and* hear while these transformations were being effected?' Since the debate between Douglas Dewar (and L. M. Davies) and J. B. S. Haldane 'Is Evolution a myth?' many fossils have been discovered which document these changes in great detail. It can now be shown clearly that the bones of the reptilian jaw joint became progressively smaller as one follows the history of the reptilian forms which evolved into the mammals—the paramammals, meaning simply 'near-mammals'.

This shrinking away of the jaw joint seems pretty inexplicable; that it actually happened is incontrovertible. Professor Fuzz Crompton of Harvard has been able to explain the reasons for these changes. If the fossil history of the paramammals is traced, it can be seen that the tooth-bearing bone of the lower jaw gradually became larger. At the same time, the shape of the jaw gradually changed. From what we discussed in the previous chapter, it is reasonable to postulate that the shape of any bone is partially a reflection of the muscle forces acting upon it. In the primitive reptiles the jaws closed through the contraction of a single large muscle. The change, in shape of paramammalian jaws indicates that this muscle split into two major components; one lifted the jaw forwards and up, the other drew it backwards and up. The result of this development was an increase in the strength of the bite that these animals could inflict. From Crompton's study of the mechanics of this arrangement, it is evident that the force exerted at the joint itself lessened. Any part of the skeleton not much used will eventually wither away. This is exactly what is thought to have happened to the two bones of the joint. In contrast to this, the tooth-bearing bone increased to such an extent that it eventually made contact with the temporal bone of the skull, where it remains in present species.

Once this new articulation had been established, the bones of the old joint became redundant. The problem, however, remained as to how any why these bones arrived in the inner ear. In the last few years Dr Jim Hopson of Chicago has provided some answers. The main sound-conducting bone, the stapes, was in contact with the quadrate—a hangover from our fishy ancestry that had somehow never been lost. The quadrate also maintained its connection with the articular. The key to the next development concerned the articular, which had a somewhat forward-projecting prong. What seems to have occurred is that this prong came into contact with the ear drum. Once this happened, sound waves hitting the ear drum would not only be carried to the inner ear by the stapes, but would also travel from articular to quadrate

The tooth-bearing bone, the dentary, increased in size and the bones of the jaw joint shrank.

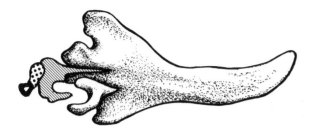

In the development of the embryo, the ear bones are initially associated with the jaws.

to stapes. This route, although longer, was more efficient, since at each step the sound would be amplified. Hearing would become much more acute—a considerable survival advantage. The direct connection between the stapes and ear drum was eventually lost and the malleus-incus-stapes mechanism was gradually established.

The possession of three ear bones is the hall-mark of a mammal is considered the most important distinction with regard to the skeleton. An accelerated version of the evolution of the earbone system can be seen during the development of the vertebrate embryo. It is most clearly seen in the more primitive mammals that begin life with reptilian, or at least paramammalian, jaws.

Another odd bone, which like the ear bones does not articulate with any other part of the skeleton but survives in glorious isolation, is the penis bone. Many a young woman on her first acquaintance with an erect penis has been convinced that it contained a bone. Every man must at some time have wished that he had been endowed with such a useful personal bone. He knows that he does not possess such an asset and his mate is never very long in making the same discovery. Of course, this does not prevent penis bones being the subject of bawdy jokes—indeed penis bones have a real existence in our folklore.

As a consequence of our personal experience and the reticence of one of the world's foremost authorities on bones, the notion lingers that penis bones only exist in the mind. In Professor Flower's standard work on mammalian osteology (published during the latter part of the last century and recently reissued), one can search in vain for an account of this bone. Few people are aware that our lack of a penis bone is more the exception than the rule, especially among our closest evolutionary relatives.

Penis bones are known in the following mammalian groups: whales and dolphins (Cetacea), walruses and seals (Pinnepedia), dogs and cats (Carnivora), bats (Chiroptera), rats, mice and squirrels (Rodentia), hedgehogs (Insectivora), bush-babies, monkeys and apes (Primates). Notable exceptions among the mammals are hyaenas and man alone among the primates.

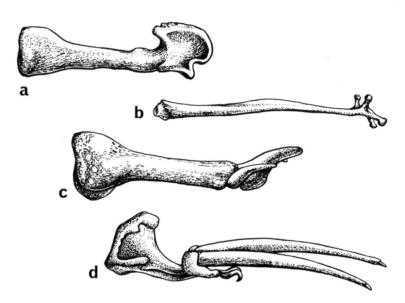

Penis bones: (a) squirrel, (b) carnivore, (c) vole, (d) jerboa.

The major part of the penis consists of the two corpora cavernosa through which the urethra runs enclosed by an extension of the tissue of the glans, the corpus spongiosum. The penis bone is associated with this last tissue and in most instances partially encloses the urethra. The dog's has a deep groove through which the urethra runs. The tip of this bone is quite simple. Some specimens, such as that of the walrus, are massive and do not have any urethral groove. Dr John Attridge has in his collection a walrus penis bone that fractured during life and subsequently healed. This may reflect the energetic love life of the walrus, or perhaps merely the enthusiasm of that one particular individual! In many animals the tip of the penis bone is expanded into bizarre shapes. One mammal called *Cercoleptes* has a penis bone sporting four small spikes and round knobs. A squirrel's has the tip flattened from side to side forming an expanded vertical plate. Other rodents, such as *Jaculus* of the mouse family and the vole, have claw-like prongs at the bone's extremities.

Since females possess as part of their sexual equipment the true homologue of the penis, the clitoris, one wonders whether any clitoris bones exist. They do indeed; but not so commonly.

Clitoris bones are generally extremely small, as is perhaps to be expected. Raccoons, the cat family, walruses and seals, rodents and primates all boast possession of these tiny clitoris bones.

The distribution of penis bones among the mammals strongly suggests that at some point in our evolutionary history our own ancestors possessed such a bone. It follows, therefore, that somewhere along the line we must have lost it. In view of our present numbers, it is unlikely to be a case of atrophy by disuse. According to Desmond Morris in *The Naked Ape*, man is distinguished by his comparatively large organ. Man's penis habitually hangs down, a feature of the primates who alone among mammals have a penis free from the body wall. For a tree-living creature and especially a vertical biped, like man, there is obvious advantage in having an excretory organ freely influenced by the pull of gravity. It is quite important for the penis to be pendent during micturition, for if it occurred with the penis still held into the body wall, the muscles at the base, which are responsible for squeezing out the last few drops of urine, would have difficulty in functioning adequately. So in the interests of basic hygiene, a vertical biped is obliged to have a pendent penis.

A free-hanging penis containing a bone would constitute a very considerable fracture hazard for the individual! Its loss is the evolutionary price we have had to pay for having a large penis and a bipedal gait. This is one of the skeletal attributes we have had to sacrifice for the sake of evolutionary progress. Many a man must have cause to regret that through evolution this bone has been lost; it must seem to many that something went wrong with evolution!

In the next chapter we shall examine what can go wrong with bones that do not disappear.

5 Bones and disease

Bones can tell us something about the way of life of people and frequently the manner of their death. For some reason or other, there is a general belief that people in the past were healthier than they are now, and that many of the diseases from which we suffer are a direct consequence of our advanced civilisation and current mode of life. However, when we look at the bones of prehistoric man, we find that he was just as much prone to disease as we are.

The skull of Rhodesian man, between 30 and 40 thousand years old, figures in very many books concerned with the history of man. This particular skull has a neat hole on the left hand side of the temple bone. Peter Kolosimo in his book *Not of this World* (1969) claims that this 'seems to demonstrate clearly the hole made by a bullet'. He proffers this as evidence of 'earlier cosmic encounters' on this planet. Unfortunately this author must have been entirely ignorant of diseased skulls. As well as the hole in the side of the head there are holes in the bone around the roots of the teeth. They were caused by dental abscesses. Furthermore the teeth themselves were astonishingly rotten. The abscess on the side of the head similarly eroded away the bone to accommodate its growth. It must eventually have severed an artery and was most probably the direct cause of this particular individual's death.

Dr Calvin Wells in his book *Bones, Bodies and Disease* (1964) described Rhodesian man in the following terms: 'As we gaze in wonder at the rugged strength of the Rhodesian man, awed by the craggy buttresses of his brow ridges and the massive sweep of his muzzle, our imagination is slow to see the end of this strange creature; to see him, long accustomed to the pain of rotting teeth and dental abscesses, now maddened by the agony of acute mastoid infection, enfeebled by toxins and the mounting pressure of pus within the bone, pus which has already plunged into the tissues of his neck and even tracked into the cavity of his cranium, bemusing him with headaches, throbbing vertiginously at every toss of the great skull; a crazed and pitiable object, panting with fever as his last hours crept slowly away, unattended, unassuaged'.

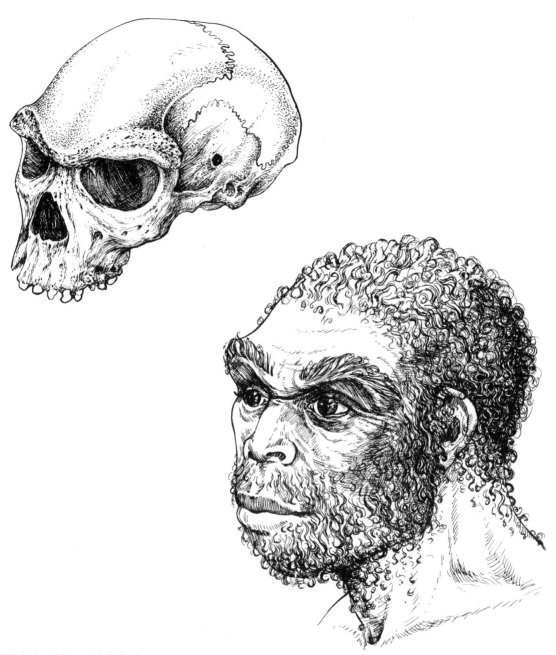

Rhodesian Man with holes by the roots of his teeth and in the side of his head caused by abcesses.

This doesn't sound as if Rhodesian man was an example of a healthy, rugged individual!

But this is where we meet a very important problem. Normally, the people who study fossils examine and describe the remains of skeletons. They are particularly well equipped for this task but they do not seem to have much familiarity with the pathology of bones. They can recognise healthy bones, and they can describe minute differences observed in bones in astonishing detail, but their accounts tend to be purely descriptive.

One of the consequences has been the entrenchment in the literature of concepts of prehistoric man so inaccurate as to verge on the ludicrous. The classic example of this is Neanderthal man, who is sadly depicted in virtually every book on ancient man as a brutish and shambling ape-like creature with his head thrust forward, very short bull neck, pigeon-toed and bent. The stance and gait of Neanderthal man is reminiscent of an ape's. This view fits well into one of our basic psychological concepts of early man —that he *should* look brutish and ugly; and in all the reconstructions Neanderthal man is certainly that. This familiar reconstruction is based essentially on a complete skeleton found in France and described by French scientists, a description subsequently accepted by generations of people. More recently however, it has been re-examined by medical men, in particular Professor A. J. Cave, and they have discovered that there was something abnormal about this skeleton. It was diseased from top to toe. The skeleton was that of an old man between *sixty* and *seventy* years old, at a time, let it be remembered, when the average expectation of life for man was about *thirty* years! (Incidentally, for a woman it was about twenty-three years—the latter being a consequence of yearly pregnancies from about the one of 13 so that by 23 she was finished). To return to Neanderthal man: he was very old, and lost all his teeth except for a few at the front and his entire neck and most of his backbone was severely distorted by osteoarthritis. When he walked it would have been with difficulty and, moreover, his gait must have approximated to the shambling one generally depicted. It does, however, seem a little unkind to attribute the stature of this poor, diseased old man to every other Neanderthal.

This example emphasises the importance of being able to recognise diseased skeletons. There are two reasons for this: first to avoid making the sort of mistakes that have already been made in the past, and second, in many ways more important, to learn about early man and his way of life from studying the injuries which he suffered, injuries which in some instances killed him, others which he survived, and also the diseases to which he succumbed.

It is equally important to be able to recognise a normally preserved skeleton. A classic case is the little dinosaur preserved in the 150 million-year-old Lithographic Limestone of Bavaria. It has been claimed that this animal died of tetanus or lock-jaw. The arching of the back and the way the head is twisted is certainly reminiscent of a tetanic spasm. In fact it is no such thing, rather it is the simple consequence of the rotting away of the flesh and the subsequent shrinking of the tendons and ligaments. The appearance of tetanus is purely fortuitous. This example emphasises the pitfalls of uncritically applying medical knowledge to ancient fossils!

Small dinosaur preserved, with its head and neck twisted backwards, giving a false impression of tetanus.

Man has been defined as 'Man—the weapon maker', and if we look at the skeletal remains of early man we see that all too often he met his death very suddenly, very violently and at the hands of his own kind. From almost the very beginnings of human history we find evidence of man meeting a violent end. Among the remains of ape-men, the Australopithecines, from South Africa, there are skulls with very characteristic double fractures. These were clearly made by the wielding of the long leg bone of an antelope. We could catalogue hundreds of pages of examples of the violent end of prehistoric men, but we will be content with just a few.

A site in China, about twelve thousand years old, preserves a family: a man with the left temple staved in, a woman with a spear wound on the top of her head, and three children, one newborn, which have been clubbed to death. At a Bavarian site, some five thousand years old, were found 33 skulls of women and children, the heads packed like eggs in a basket, all of them with the roofs of their skulls smashed in by some sort of stone mallet. Coming nearer to the present time, from a site in Brittany there is a very famous case of a stone arrow head sticking into a vertebra of a skeleton. There is no sign of any reaction of the bony tissue to this object. This arrow head must have severed a major blood vessel and been the direct cause of death. Occasionally these sort of wounds and fractures healed and the victims survived the ordeal.

A wound in the thigh of early man from Java, the 500,000 year old *Homo erectus*, tells us little about how it was received. Near the hip joint there is an irregular proliferation of bone which has been the subject of much argument; it has even been described as an example of a prehistoric cancer. What seems much more likely is that this particular individual received a deep wound, which led to profuse bleeding. When a large blood clot forms internally, it

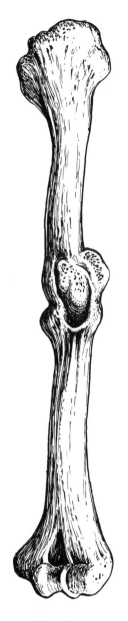

*The upper arm of
Dr David Livingstone showing
the healed fracture.*

frequently becomes calcified and the whole clot is transformed into an irregular mass of bone. This is what may have happened to Java man a long time before he actually died.

In 1361 at Visby in the island of Gotland, there was a major massacre of the inhabitants by invading Danes. The skeletons of the Gotlanders show sword cuts on their legs, very few on their arms, and these sword cuts were always on the left side of the body. This suggests that the Gotlanders were well armed in the upper reaches of their bodies but that the right-handed Danes had long swords with which to slash the legs of the defenders. There were also strange multiple penetrations of the Gotlanders' skulls which seem to have been caused by objects known as 'Morning Stars' wooden balls with metal spikes in them, attached by a chain to a wooden handle. These were whirled round the assailant's head and brought down with considerable viciousness on the heads of the victims, leading to instantaneous death. These few examples tell us something about the history of war technology.

Naturally, not all such violent injuries need be occasioned by one's fellow man. In 1843 Dr David Livingstone was attacked and mauled by a lion during which his humerus suffered a compound fracture. A later injury to the same arm gave him a 'false joint' in his upper arm. When Livingstone's remains landed at Southampton on 15 April 1874 the post-mortem examination that same evening showed the fracture callus and 'false joint'—ample and wholly convincing confirmation of the identity of the remains.

Violence of the big cats has in the past also been turned against other cats. One of the most striking examples is the prehistoric cat *Nimravus* which had an encounter with a sabre-tooth tiger. The long dagger-like stabbing canine had plunged into the muzzle of *Nimravus* destroying the nasal passages on one side and even the olfactory lobes of the brain. This traumatic event did not lead to the immediate demise of the victim, and the wound subsequently healed. One cannot but wonder what happened to the sabre-tooth tiger in this affair!

All these deliberate and violent attacks on individuals are not the whole story. Man does not live by war alone; he indulges in other, more pacific, pursuits which even so can entail violence to his person. These may give us interesting insights into the occupations of the people concerned.

For example, the shin bones of the Anglo-Saxons are frequently fractured. Twenty per cent of Anglo-Saxon remains show evidence of a characteristic fracture which is caused by the foot being very suddenly twisted on the leg, a fracture due to methods used in breaking and cropping rough ground. Furthermore, there are many fractures of the forearm about an inch above the wrist, which may have been caused by falling on to an outstretched hand, a fact which tells us something of the farming of the Anglo-Saxons! In contrast, in Egypt, 10 per cent of fractures are in the legs but 30 per cent in the forearms, the latter being in the middle of the forearm on the outer side, and are parry fractures which are the result of blows. The forearm was raised above the head to protect it and when the blow fell the forearm was broken. This type of fracture is particularly frequent in women, telling us something about the relationships of the sexes! It suggests that wife-beating was a common activity. There is one rather poignant case of a teenage girl who put up one arm to protect herself, her fore-arm was smashed, she put up the other and this too was smashed and then finally her skull was staved in. This particular girl was four to five months pregnant at the time and it leaves little to the imagination to work out what occasioned this violent assault.

Another series of fractures which provides an insight into the way of life of a people, and the division of labour between the sexes, is illustrated by some Indians from California. Among the remains of these Indians are very many fractures of the lower limb bones in the men and none at all in the women. This particular tribe lived by the seashore and collected food from the sea; the men spent their time on slippery rocks, whereas the women collected from rock pools. Hence, the men were always breaking their legs, the women hardly ever.

So far, all the injuries inflicted have been accidental, at least from the victim's standpoint. But one particular lesion is especially noteworthy. It was received at the Battle of Prestonpans in 1745 by Captain Clarke Morris, who on being unhorsed was struck by a claymore 'which sliced off the flesh and a piece of bone from his skull'. He recovered from this and lived for many years with a silver plate over the wound. It is not known when he died but he was reported to be in Lisbon during the Great Earthquake of 1755.

It is thus evident that people can live with pieces of skull missing, indeed the surgical removal of discs of bone from the top of the skull has been practised for many centuries. Perhaps the most famous, or notorious, are the trephined skulls from the Andean countries of South America, notably Peru. This delicate operation was performed with primitive stone implements and some indi-viduals survived repeated trephinations. It is not known whether such drastic procedures ever aided the patient—one can merely surmise that 'evil spirits' were released.

In parenthesis, it may be mentioned that there are recent accounts in the press of men in England suffering from headaches (and also being somewhat under the influence of alcohol) and who persuade their friends to hammer nails into their skulls to cure their headaches! Recently a man was reported having committed suicide by hammering seven six-inch nails into his head. Readers of the Bible may recall the account in Judges chapter 4 verse 21 of the manner in which revenge was taken on Sisera, oppressor of the children of Israel: 'Then Jael, Heber's wife, took a nail of the tent, and took a hammer in her hand, and went softly into him [Sisera] and smote the nail into his temples, and fastened it into the ground: for he was fast asleep and weary. So he died'. No wonder!

Another method of interfering deliberately with man's bodily frame is to deform it, in the interests of supposed beauty. The notion of what is desirable in the shape of a head varies from one culture to another. Some cultures are not content to let Nature take its course but take active steps to ensure the desired results.

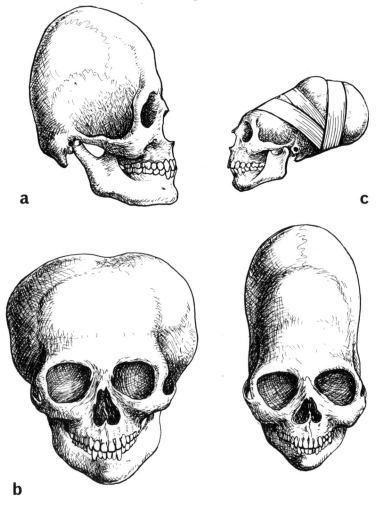

Artificially deformed skulls:
(a) the back of the head flattened by a cradle board, Navajo tribe, Arizona;
(b) flattened by a board, Peru;
(c) elongated by binding, Peru.

In North and South America skull deformation was accomplished by the baby's head being bound to cradle boards. The confined space resulted in the developing skull altering its proportions to accommodate the growing brain. Some examples of these are shown in the accompanying figures. Some of the more spectacular skull deformities are the result of bandaging the head, a former practice in parts of Africa and Peru. Reluctantly, we have to admit that the resulting elongated skulls do have a rather appealing elegance. Indeed many hairstyles and head-dresses appear to simulate these very proportions.

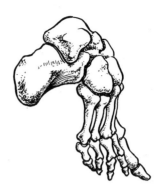

Bound foot of Chinese woman.

Again we are shocked when we read of the Chinese practice of binding girls' feet. A lady was not intended to work in the fields but rather to be a delicate flower child. This practice was cruel because the way of life for which such girls were prepared was going out of style. Women whose feet had been bound during childhood and who by cruel conditions were subsequently denied a life of leisure, were obliged to carry out their work hobbling painfully in their permanently deformed feet. The Chinese woman's foot we illustrate shows the crowding of the toes and the exaggerated arch of the foot as a result of forcing the heel forwards and down. In spite of its unnaturalness, it is not difficult to understand the appeal of such a shape.

Deformities less contrived but of a more debilitating nature develop of their own accord. These are slow accumulative diseases usually associated with ageing, of which the most familiar is arthritis. The lubricating synovial fluid within arthritic joints somehow deteriorates and the cartilaginous capping of the articulating surfaces of the bones is plucked off, as mentioned in Chapter 3. Eventually bone grates directly against bone, and in this condition the joint wears away with the accompaniment of great pain. The bones become grooved and highly polished like ivory—indeed, the process is termed eburnation. An example of this condition is shown in the illustration of the arthritic knee joint.

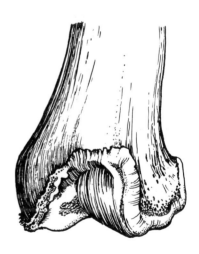

Arthritis of knee, with bone surface ground away and proliferation of bone around the edge.

The grinding away of the bones in the joint stimulates a reaction, the adjacent areas not subjected to direct wear experiencing a reactive proliferation of bone. In the early stages this extra bone forms a lip around the edges of the joint, a process termed lipping or osteophytosis.

Unless the joint is kept moving the proliferating bone from both sides will meet up resulting in a complete seizure of the joint. A rather dramatic example of this is shown in the knee joint of the amputated limb of a 45 year old man. The entire joint is fused or anchylosed, the ligaments of the joint are ossified and the main bones are joined by columns of new bone. Even the knee-cap is fused to the thigh bone. The rather attractive spiky ornament is due to irregular deposits of new bone.

In old people the hands frequently become arthritic, and the wrist bones fuse as do the metacarpals of the palm, the ends of these bones proliferating to give the characteristic knobbly appearance to the hand.

In modern Europe arthritis is extremely common in the vertebrae of the lower back, especially the fifth lumbar. This, it is said, is due to the perpendicular thrust of the weight of the body trunk, which compresses the discs, accentuated by spending too much time in lounge chairs, in cars and slumped over desks.

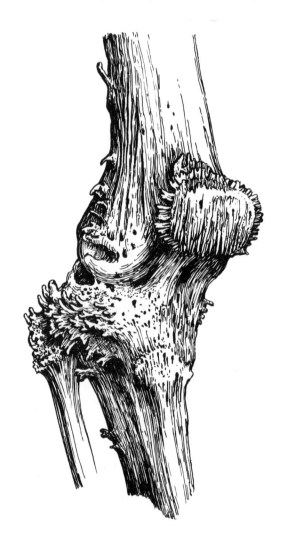

Arthritis of knee, with entire joint fused.

Arthritis of the fifth lumbar vertebra was common in Anglo-Saxons too, and has been attributed to tilling hard ground, though whether this was so or not is difficult to tell. It may well be that the vigorous pelvic thrusts which accompany our sexual activity may be a contributory cause of this particular arthritis. It is certainly true that after a vigorous weekend one has a most dreadful pain in the small of one's back!

The number of arthrites in odd parts of the skeleton may indicate specialised activities. For example, the Patagonians who used to hunt with the bolas developed arthritis in the shoulder and elbow joints. The Minoan bull-leapers developed osteophytosis of the spine. Neanderthal man, at least one in particular, became extremely arthritic but this could be a general feature of old age.

The cave bears which suffered arthritis in their spines were, without doubt, basically suffering from the ravages old age. This was also the case with the giant plesiosaurs, which show typical osteoarthritic lipping of their vertebrae. When one of the present authors first noted this lipping, we described it in great detail; only years later did we realise it was simply arthritis. This emphasises once again the need for students of fossils to be aware of the main features of bone diseases.

Arthritic vertebra of an extinct Cave Bear, showing lipping of bone around the margin.

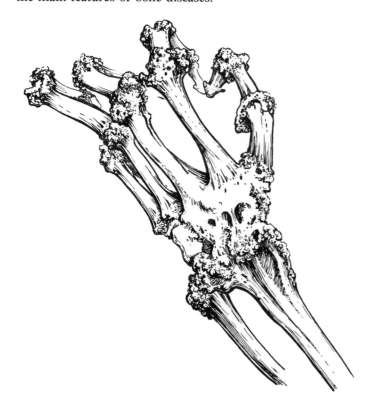

Arthritic hand.

At the same time there is an inherent danger in letting bone pathologists loose among fossils. In most giant dinosaurs there are two vertebrae in the middle of the back that are fused. These have been identified as typical examples of osteoarthritis by a medical colleague. They are no such thing. These two vertebrae are fused for strength, as they are a kind of keystone of the entire backbone.

One thing, however, is clear and that is that osteoarthritis goes back in time for several hundred million years. It is a degenerative disease of enormous antiquity—little consolation if one happens to be suffering from it.

As well as degenerative diseases there are a whole series of infections that can affect bone, the commonest of which is periostitis, which gives a fluted surface to the bone as a result of

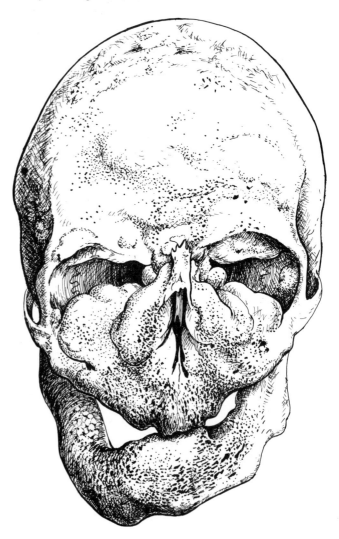

Leontiasis (lion head), a disease caused by proliferation of the bones of the face.

the inflammation of the tissue that covers the bone, the periosteum. If the infection of the bone surface continues, so that the marrow of the bone is invaded and becomes filled with abscesses, the condition is then called osteomyelitis. It is especially common on the shin bones of Anglo-Saxon skeletons, again due to their falling over and breaking the skin; with soil and dirt in it, a graze would quickly become infected. Surprisingly, in the ancient Egyptians, there are similar sorts of lesions on the shin bones even though there were few fractures of the lower limb. This condition can be explained in terms of insect-infected sores.

A disease said to be a nasal infection, but which is still not properly understood, is leontiasis or 'lion-head'. In this disease the bones of the face proliferate, gradually closing up the eyes and nose, and produce a grotesque appearance—perhaps the worst feature of this condition. We have illustrated a fairly typical example from Peru, and this is by no means an extreme case.

Probably the most horrific looking example is that of a man who died at 34. He noticed at the age of 14 that the bones of his face were enlarging and, apart from the unsightliness, he suffered little inconvenience until his last two years when the pain became intense. His eyeballs protruded almost beyond the lids and the sight in one eye was lost, yet his senses of hearing and smell were unaffected and his intellect was unimpaired. The appearance of this skull is reminiscent of a clipped poodle dog and not of anything human.

A one-time common deficiency disease which affected the skeleton was rickets. This is essentially due to a defect in calcium metabolism with insufficient calcium being deposited in the developing bones, leading to buckling under the weight of the body. The skeleton of a 70 year-old woman who suffered from rickets as a child illustrates this condition: the thigh bones are curved forwards and the shin bones are curved forwards and outwards—in fact, the whole posture is distorted. The hip joint is displaced and the pelvis as a whole is flattened and contracted. Milk, vitamin D and plenty of sunshine are the simple elements preventing this disease.

Another ailment which affects the bones of young children is infantile scurvy, which is due to a vitamin deficiency—the child being deprived of natural milk and being fed prepared foods in which the vitamins have been destroyed by heating. The characteristic appearance of rounded bosses of spongy bone on the skull is shown in the illustration of a 9-month-old child.

One of the most emotive diseases, and the one about which there is perhaps the greatest publicity, is syphilis, transmitted by sexual

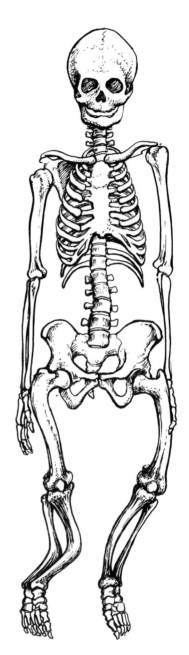

Skeleton of a 70-year old woman, who suffered from rickets as a child.

51

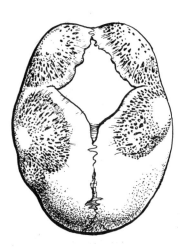

Top view of the skull of a 9-month old child, suffering from infantile scurvy.

activity. It is only in its last phases that this disease affects the skeleton but it does so in quite a dramatic way. Eroding ulcers form on the skull and long bones accompanied by a reactive proliferation which gives a very characteristic appearance to, for example, the skull. Many areas of bone become completely dead or necrotic and, accompanying the thickening, there are fistulous openings through which pus and decaying matter exude.

Syphilis is caused by the spirochaete organism, *Treponema*, which has a considerable history. It is possible, from the work of the anthropologist Don Brothwell, to trace the likely evolution and spread of the diseases caused by this organism. There are four related diseases: pinta, yaws, endemic syphilis and venereal syphilis. It is now believed that pinta was world-wide by about 15,000 B.C. and was subsequently isolated in the Americas. From about 10,000 B.C. it is believed that the voyages of the Amerindian peoples across the Pacific carried a new mutant of *Treponema* which gave rise to the disease yaws. This became firmly established in the Pacific area, south-east Asia and India. Yaws differs from pinta in that it affects bone, causing a series of inflammatory changes which involve the erosion and destruction of the bone, with the pitting and slight proliferation characteristic of advanced venereal syphilis. The nasal region begins to rot away and the front incisors tend to drop out. Furthermore, one finds openings in limb bones where the infection is invading the skeletal bones. By about 7,000 B.C. it seems that there was a further mutational change resulting in the appearance of endemic syphilis. This first developed in Asia and seems to have spread through the agency of Genghis Khan's hordes into Asia Minor. And it is in this latter region, at about 3,000 B.C., that venereal syphilis appeared in the cities of the civilisations of this area. During the 15th century an exceedingly virulent form of venereal syphilis spread into Europe. In Europe and Africa there was no evidence of either yaws or syphilis before about A.D. 1500. In fact the spread of syphilis into Europe and Africa seems to be correlated with the expansion of the Islamic world. The pattern of the spread of venereal syphilis just outlined, contrasts very much with what is normally assumed to have been the case. The great epidemics of syphilis which took place shortly after Columbus and his sailors had returned from the Americas have led to the belief that they had acquired this disease from the Americas and had introduced it from there into Europe. The belief in the American origin of syphilis dies very hard and it seems to have been due to the confusion of yaws with venereal syphilis. In fact, as Calvin Wells said, 'to accept the Columbian origin for the disease commits us

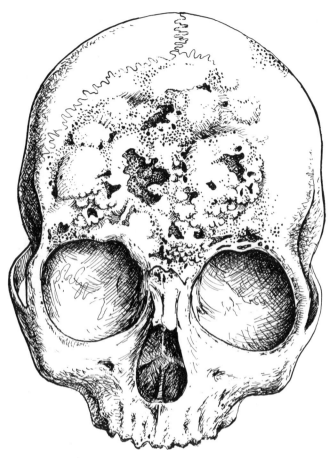

Syphilis.

to a belief that less than fifty sailors were sufficient nucleus to infect the continent of Europe within eighteen months'.

Venereal syphilis, or something very like it, seems to have existed in the Americas before the time of Columbus but its origin seems to have been in Asia. We can be quite certain that syphilis was not introduced into the Americas initially from Europe; it was already there before Columbus arrived. And the final sobering thought on the history of syphilis: with the introduction of antibiotics during the last quarter of a century, new resistant strains have developed and these have already spread around the world within this short period so that we are now experiencing a new syphilitic epidemic.

There remains one final topic to discuss in this chapter: the ovarian cyst. Such cysts and their bizarre contents have had a fascination for medical men for generations. Some contain skin, hair and sebaceous oils and fat, other have numerous teeth in them and also bones. The teeth that occur in ovarian cysts are clearly well-formed, and one can identify incisors, canines, and molars. At the same time, isolated mandibles, maxillary bones and other skull elements can be recognised. In fact, we have seen a description of an 18-year-old girl in the earlier part of this century, who had a very large ovarian cyst which contained 150 teeth. Such teeth are well known and Professor Richard Owen described microscopic sections of them in the 1840s. Occasionally, it has been recorded that teeth develop in the testes, but this phenomenon is exceedingly rare in man. It is, however, fairly common in horses which seem to be prone to this sort of development. Ovarian cysts are believed to arise parthenogenetically (by virgin birth) from the germ cells. They are, in fact, disorganised foetal structures which are the product of a kind of virgin-birth. All contain skin structures, and 90% contain teeth and bone.

With regard to the ovarian teeth, it is tempting to speculate that the ancient myth of the female of our species having teeth in her vagina may well have arisen from the post-mortem discovery of a woman with teeth in her ovary. One presumes that the contraction of the strong vaginal sphincter muscles could quite easily give rise to the idea that the vagina had the potential of detaching the penis from its owner. It is not difficult to imagine the effect on a primitive people of the discovery of ovarian teeth. In their eyes this would most certainly confirm the suspicion that 'genuine' teeth are able to perform such a ghastly operation.

The uses 6
of bone

The history of civilisation is the history of man's technology. It began with bones when he became a meat eater and learned to trap and hunt animals. When he had eaten his fill all that he was left with were bare bones.

A somewhat imaginative account of the beginnings of technology was shown in the opening scene of the film *2001*. Here were a group of ape-like creatures, one of which happened to be playing with the skeleton of a dead zebra. He hit one of the bones, which flew into the air. Grabbing it, he realised that he had in his hand a weapon, an extension of the hand, and this is essentially what tools and weapons are. Indeed, it is not at all difficult to envisage man beginning his long uphill struggle to the present from the point at which he learned to wield bones.

In fossil deposits, a million or so years old, in southern Africa, are found the skulls of baboons and also of ape-men, the Australopithecines. A high proportion of them have been caved-in with an object which imparted a characteristically double fracture. Such a fracture could only have been made by the articular surface of the upper bones of the fore-legs of antelopes and other herbivores. It has even been deduced that the fractures in the skulls of baboons and Australopithecines were made by an animal which was right handed! It seems perfectly reasonable to infer, since we know of no other organism that has this propensity for weapon wielding, that these baboons and also some early men met their ends at the hands of beings wielding the long-bones of what were once their meals. The acquisition of such weapons must have given an enormous advantage to these first men, such is the opportunism of mankind. Even Samson found the jaw-bone of an ass of considerable value in his activities.

However, man is defined not so much as a weapon *wielder* or tool user, but as man the weapon *maker*, involving the modification of natural objects. Such a course of action implies that the people concerned were capable of consciously planning ahead and producing tools for future contingencies, a trait distinguishing mankind from the rest of the animal world. In the deposits containing the smashed skulls, there is an enormous variety of bones, generally fragmented. Professor Raymond Dart has proposed that many of these bones were in fact tools fashioned by primitive man. Dart coined the word 'osteodontokeratic', which simply means bone, tooth and horn, to describe this culture. He cites the distal ends of long-bones, especially the humerus, which are broken to produce a dagger-like spike, and which he contended was a digging tool or dagger. He noted that there were ten times as many distal ends of the bones as there were proximal, suggesting some sort of selection.

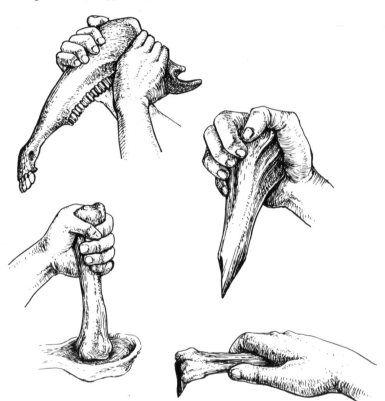

The use of bones as tools by the earliest men (after Professor R. Dart).

Dr Bob Brain studied the living Bushman of the Kalahari Desert and their food. He examined the goat bones scattered around the camps and discovered that only the distal ends of the humeri were present, never any of the proximal ends! Once the people had eaten all the meat, they would split the bones for the marrow, what was left being thrown on the ground for small dogs to chew and gnaw. The interesting feature of this study was that all these distal ends of bones had the identical shape of the supposed digging or stabbing tools described by Dart! They had broken in the same sort of way, and ended up as spiky chunks of bone.

Dart also described numerous polished points and flakes of bone. These too, Brain suggested, were parts of discarded bones. Bone fragments tend to collect at water-holes and the tramping of both people and animals to and fro leads to the splinters of bone being polished by the sands in which they lie. So here again there is evidence that some of the so-called tools described by Dart can be explained as natural phenomena and do not necessarily reflect tool-making activities of early man.

The accumulations of bones described by Dart have been variously interpreted. It has been suggested that they represent collections made by hyaenas, a view strenuously denied by Dart, for the hyaenas of South Africa do not accumulate bones. But Dr Tony Sutcliffe has been examining hyaena dens in other parts of Africa and has proved that, there at least, hyaenas do accumulate bones, even human ones. The working on many of the bones described by Dart has been claimed to have been made by the gnawing of porcupines, and it is known that porcupines have a special relationship with hyaenas, managing to sneak off with bones to their dens to gnaw them.

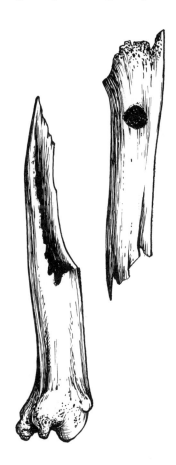

Pseudo-tools made by hyaena digestive juices and gnawing.

The concept of the 'osteodontokeratic culture' has been attacked with incredible vehemence by many anthropologists and palaeontologists. It is a little difficult to understand why this should be so. Perhaps it is because the theory has been put forward by Professor Dart, who is noted for the enthusiastic controversies into which his work always seems to lead him. Nevertheless, it is true that many of the bones described by Dart can be explained in other ways. For example, what seems to be the most convincing sort of man-made tool, the insertion of one bone inside another, can in fact occur quite naturally, as Tony Sutcliffe has shown. When hyaenas swallow bones, they become eroded by the gastric juices, are regurgitated and quite fortuitously bones can become inserted into one another. In other cases, bones appear to have had holes drilled in their sides, again possibly the work of gastric juices.

The accumulations of bones were clearly connected with man's activities. Most were probably the remains of the animals he ate, but the crucial question is whether any of them were naturally converted into tools, for either cutting meat off bone, scraping skins or whatever. The answer seems to be an unequivocal yes! For although some of Dart's material can be otherwise explained, there is nevertheless a considerable residue which can only be explained as having been fashioned and used by man. For example, in the polished points and flakes, a genuine tool is distinguished from a pseudo-tool by the fact that its polish is confined to a single surface. Similarly, the chipping of bones to give a sharp surface, or any sort

57

of surface for that matter, is unlikely to have been produced naturally if the working is on one surface only. This indicates a directional activity by some sort of organism, in this instance, of course, man.

There has been so much emphasis on stone tools or flint instruments that it is often forgotten that bone tools and bone instruments also have a very long history. There are primitive peoples today who use bone tools, as well as stone ones. It is very difficult to understand why a bone-based culture should be denied, since bone must have been one of the most convenient, at-hand materials available.

Bone harpoons, spear heads and needles from Stone Age Man.

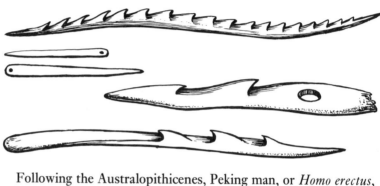

Following the Australopithicenes, Peking man, or *Homo erectus*, best known for his stone hand-axes, also used bones. Chipped bones were used as scrapers for preparing skins. Awls were fashioned as were needles complete with eye. It is important to remember that although Peking man only fashioned very crude stone implements, his bone technology was clearly much more advanced, and this itself suggests that the fashioning of objects out of bone, rather than stone, was of greater antiquity. Later in man's history, with the advent of our own species, *Homo sapiens*, organised flint mining was carried on, and deer antlers were used as picks for levering out blocks of stone, and the shoulder blades of oxen served as shovels. Parts of antlers and bones were further used for pressure flaking, one of the techniques in manufacturing the more sophisticated flint tools.

Although Stone Age man is rightly renowned for his flint instruments, he did not neglect bones. There is found a greater variety of bone tools than in previous ages, which proves that a bone culture was not only surviving, but was expanding in parallel with the flint industry. Antlers were used as hammers, others as shaft straighteners. Barbed points, fish hooks, harpoons, spear points and even a spear thrower are known to be made out of bone or antler.

Later, we reach the Neolithic Age when agriculture began: instruments were used for harvesting grain, including sickles of

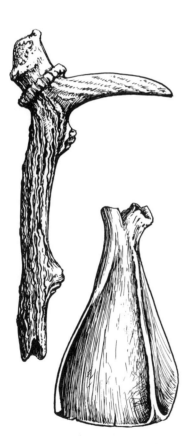

Bone pick and shovel used in flint mining by early man.

58

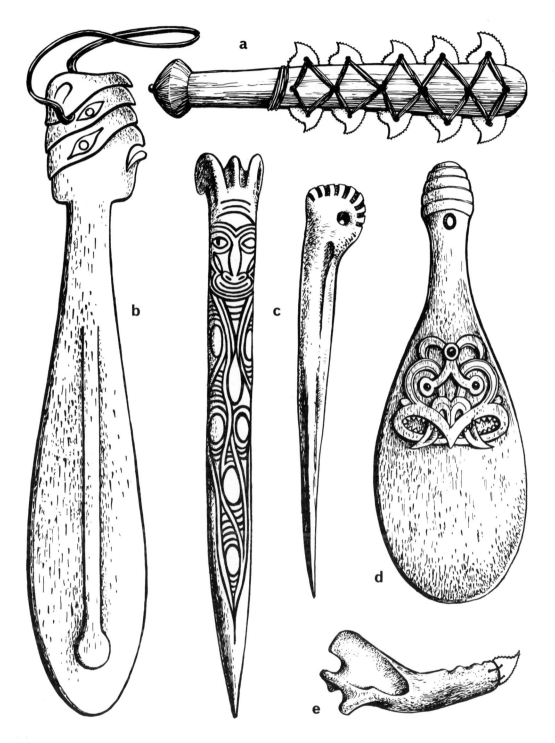

Weapons: (a) slashing knife of shark's teeth, Polynesia; (b) ceremonial club of whale rib used for killing slaves during feasts, Haida Indians; (c) daggers of cassowary bone from New Guinea; (d) Patu or single-handed club, Maori, New Zealand; (e) dagger of dog's jaw with shark tooth, Polynesia.

bones or antlers into which were inserted delicate flint teeth. Here again the flint and bone industries meet, although pottery also now makes its appearance.

Throughout recorded history bones have been used as tools or adjuncts of tools. Modern man, in different situations, has discovered almost an infinity of uses for bone. The most basic, as with the Australopithicenes, was to hit people or animals over the head! Bone clubs are used by Eskimos when sealing. The single-handed weapons used by the Maoris of New Zealand and other Pacific peoples were carved out of whale bones; these 'patu' were comparatively blunt instruments, more piercing ones being constructed out of the long bones of large birds. From New Guinea daggers made of cassowary bones would have been capable of inflicting fatal wounds with great ease. In Polynesia and other regions of the Pacific, where such materials were less plentiful, sharks' teeth were utilised for the same purposes. A small dagger described by Captain Cook was made out of the lower jaw of a dog with a shark's tooth tied onto the end. For really lethal weapons, large batons of wood with their flat edges lined with sharks' teeth must have been exceedingly effective when wielded in battle.

Hunting and fishing implements have often been made of bone. For the Eskimos, whose entire livelihood depended on such activities, virtually all their tools were of either bone or walrus ivory. Indeed bone was the only suitable material available and these people developed a true bone culture, which we shall describe in the next chapter.

Sledge and skates using jaws of cows and sheep. Centre: *sledge with horse limb bone runners.*

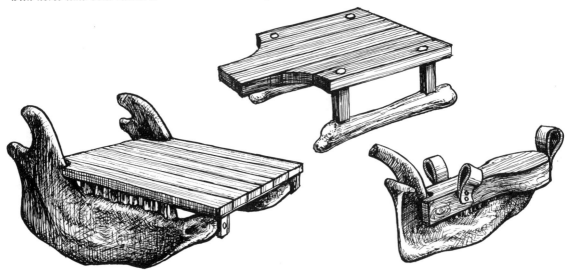

Bone has long been recognised as an ideal substance to enable man to travel over ice. In Fitz Stevens 'Description on London', written in 1180, the following account appears: 'When the Great Fenne or Moore (which watereth the walls of the citie on the North side) is frozen, many young men play on the yce . . . some tye bones to their feet and under their heeles, and shoving themselves with a little picked staffe do slide as swiftlie as a birde flyeth in the aire or an arrow out of a cross-bow'. The metatarsals of cattle were strapped to the boots by means of string or thongs threaded through holes drilled into the bone. Sometimes they may have been made out of pieces of whale rib.

In Austria and Bavaria, the cannon bones of horses were used as sledge runners. In some cases, they were pegged on to wooden stools. In north-west Poland (Pomerania) children's sledges were constructed from the jaw bones of oxen by simply fitting a wooden seat between the two mandibles. An effective way of making skates was to fit a platform of wood directly on to a sheep or goat jaw, slotting it over the teeth.

Throughout history bone has been used in the manufacture of clothing. Bone needles and awls go back to prehistoric times. The Romans used bones as shuttlecocks in weaving cloth. Bone lace is so called because the weighted bobbins are generally of bone, originally from pigs' trotters. A split bone tool is used for tatting, while fishermen made and mended their nets using bone netting needles. In the days of sail, sailmaker's seam rubbers were often of bone; so were the marlin-spikes used for splicing rope. The North American Indians who worked in bark, peeled it off trees by means of bone bark-peelers. In Denmark, in the last century, ironing boards were made from the jaw bones of whales! The small figures from Japan, netsuke, are attached to a cord which is pulled through the sash of the kimono. The weight of the netsuke prevents the cord from slipping out of the sash.

Still on the domestic scene, bones were used as the bows of bow-drills for making fire; as the bone is moved from side to side the vertical piece of wood is twirled rapidly and the resulting friction will, with patience, ignite the combustible material placed on the base block.

Numerous domestic utensils are made out of bone: spoons and scoops from every continent and every age, the decorated reindeer antler spoons being probably the most ornate. The Lapps, like the Eskimos, have a bone-based culture, in this instance exclusively based on the reindeer. Even today people consume their breakfast boiled eggs with bone spoons in preference to silver spoons which tarnish badly by reaction with the egg protein. Knife handles have

Awl for mat making, Indonesia.

Bone bark peeler, Alaskan Indian.

been made of bone from Roman times to the present; a small bone knife and fork set dug up in Charing Cross Road, London, may well have belonged to a child. The paucity of forks in collections is probably due to the fact that sharp-pointed knives served the purpose until forks proper were introduced in the 17th century.

Somewhat more macabre utensils are bowls made from human skulls. Their provenance is Tibet, where human bones seem to have been some of the more popular objects in the home. In the tropics large pieces of turtle carapaces are made into axes for chopping up breadfruits.

One of the commonest feeding tools up until this century was the bone scoop. Professor Dart found them with the Australopithecines; they are known from ancient Egypt, Rome and many other societies and are made easily from the cannon bones of sheep or goats; most of the British ones are decorated, in many cases with the initials of the owner. They were held in the hand and the working edge was scraped across the flesh of an apple, so that a pulped mush was produced which could then be eaten comfortably by either a baby or an edentulous and aged grandparent. Only through the use of this instrument could toothless old persons continue to enjoy raw apples and other hard, fibrous foods. For pastry pies, a bone jagging or crimping wheel was used to add decoration.

Still surviving on some people's dining tables are bone napkin rings. Rarely, however, does one find a household where everything that can be made out of bone actually is. For this one has to visit fanatical bone-lovers or 'osteophiliacs', among whom we number ourselves. Our friend Dr Per Persson has bone utensils in his home all of which he has made himself (he is a professional osteologist—a student of bones). The one we have illustrated is his salt cellar.

While we are on the subject of the home, let us look at one of the most unusual uses of bone. During the Upper Stone Age ancient man built his homes of bones. At that time, about 20,000 years ago, the woolly mammoth flourished in great numbers throughout the plains of northern Europe and Asia. This was during the Ice Age when tundra conditions were widespread. There is abundant evidence that in Czechoslovakia and the Ukraine man's main occupation was mammoth hunting. Indeed he must have been the major factor in their subsequent extinction. His entire life was mammoth-orientated. Evidence of his trapping and utilisation of the ivory and bone have been available for a long time. Excavations by Soviet archaeologists in the 1950s, and also in the last few years, have revealed that the mammoth bones were often preserved in an orderly manner, massive bones and skulls being arranged in a

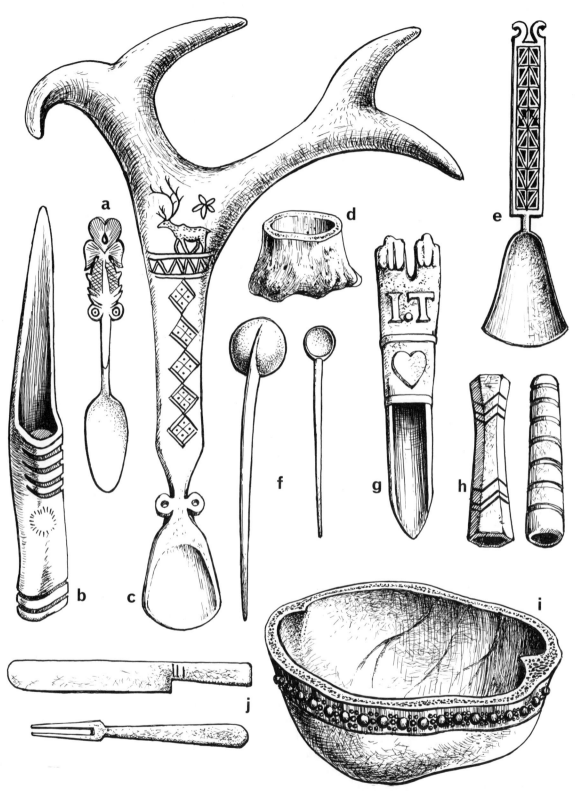

Domestic utensils: (a) spoon, China; (b) limespoon, New Guinea; (c) reindeer antler spoon, Lapland; (d) salt-cellar, Sweden; (e) spoon, Madagascar; (f) Roman spoons; (g) apple scoop, England; (h) Roman knife handles; (i) drinking bowl made from top of human skull, Tibet; (j) child's knife and fork, England.

63

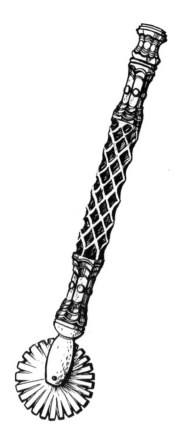

Pastry crimper, England.

circle. Skulls and lower jaws inserted into one another, together with massive limb bones, formed the foundations and outer walls of dwellings. Tusks, ribs and shoulder blades made up the walls and in some cases the roofs were built of a meshwork of interlocking reindeer antlers; in others mammoth ribs and tusks were used. Naturally, these buildings weren't entirely built of bone; the framework of the huts consisted of wooden stakes on which animal skins, particularly those of mammoths, were fixed, while the outer wall was constructed of bones, again covered with more skins. One dwelling was made up of 385 bones representing 95 mammoths, another of 400 bones from 33 individuals.

Generally there is a hearth containing ashes in the centre of these huts. The living area is about 23 square yards, the diameter being about 16 feet. Evidence of habitation, apart from hearths, is suggested by the abundance of tools, flint instruments, reindeer antler hammers, awls, bone needles, ivory spears, patterned bones and figurines and the remains of game animals. The mammoth, and especially the younger individuals, were the main object of the hunt, with the hare coming second, the skin of the latter being especially favoured as clothing material.

So far we have only been considering the utilitarian aspects of the use of bone. Let us remember, however, that man's existence is not merely a question of food and shelter: there are other areas of activity in which he engages and in these, too, bone has its uses.

Mans leisure activities in most societies includes music of some description or other. Bone or ivory clappers are one of the simplest of musical instruments, and long bones of animals are ideal for making flutes—single, double or even nose flutes. Bird bones seem to be favoured, although the South American nose flute which we

Dwelling built of mammoth bones by prehistoric mammoth hunters in the Ukraine.

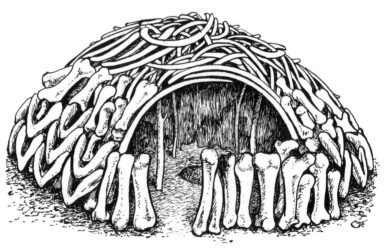

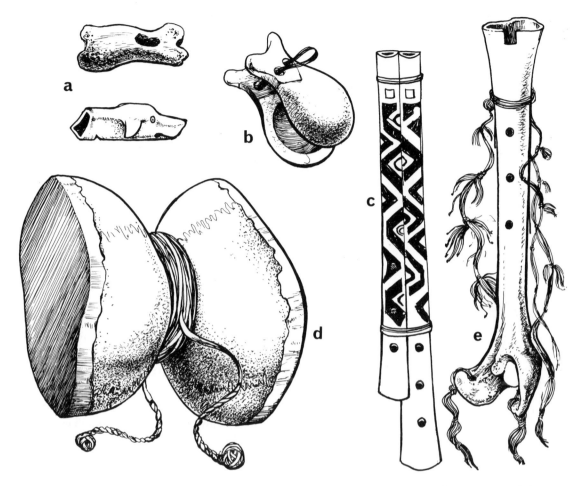

illustrate is made out of the bone of a jaguar. In Tibet trumpets of human thigh bones seem to be rather popular—the head of the bone being sawn off and a hole drilled in each of the two articulating condyles of the knee joint. An even more macabre musical instrument is the Tibetan double-membrane drum for which a couple of human skulls are needed; two skull caps with a membrane stretched across each are bound together and the drum is then grasped in the middle and twisted sharply so that two cords with pellets at their ends swing on to the membranes. These drums are used mainly by begging monks while chanting. In contrast to this, there are 17th century English dog-whistles that, on account of their pitch, can be properly appreciated only by dogs—the one illustrated is carved appropriately in the shape of a dog's head. Cruder whistles are known from pre-historic times.

In Estonia, ducks are attracted by the rubbing of two bones together; one of them is serrated and the resulting noise effectively simulates a duck call.

Musical instruments: (a) prehistoric whistle and contemporary English dog whistle; (b) clappers; (c) double flute of bird bones, Peru; (d) double membrane drum of human skulls, Tibet; (e) nose flute of jaguar bone, British Guiana.

Bone flutes used in divining, Panama.

Other pastimes are games of chance and games of skill. Bone dice are known from all ages, in many communities unmodified bones are used as dice, especially astralagus bones from the heels of sheep or goats. Gambling counters are cut out of flat bones such as sheep scapula or they may, as in the case of Eskimo counters, be carved out of ivory in the form of animals. Only *in extremis*, as we relate in a later chapter, are, for example, playing cards carved out of bone. Of games of skill, chess and draughts are the most familiar, and the pieces for these games are frequently carved out of bone or ivory. Some of the most attractive examples come from 11th century Scotland. The famous chess set discovered at Uig in Lewis is carved out of walrus ivory, as is the draughtsman we have illustrated. The draughtsmen are decorated with patterns and figures, in some cases with strange little beasts.

Children's games are also represented in bones, some of the more imaginative coming, as is to be expected, from the Eskimos who play a game of skill, similar to our pin and ball, consisting of a bone spike and one, or as many as five, hollow cones of bone. The cones are thrown in the air and the object is to catch as many as possible on the bone spike. For the younger child, there are the 'nodding' or 'pecking' birds—a variation of the eastern European toy; in the Eskimo version the birds peck alternately.

Perhaps the most ubiquitous of all toys is the doll, which may be made of every conceivable material including bone. The simplest dolls are of little knuckle or finger bones with faces painted on them and little skirts tied round. In the Matto Grosso region of Brazil, dolls are made from cow foot-bones decorated only with pigtails and wax breasts. A more elaborate example is the Roman doll with pegged arms which comes from 3rd century Tarentum, Italy.

The one game that must be discussed in any book on bones is of course 'bones'. There are many variations of this game but the actual bones used are sheep, goat or small deer astralagi—usually five; 'five-stones' is the same game, and was played by the ancient Greeks and the Romans who cast astralagi in lead and bronze. Astralagi have been used for these games throughout the world, with the exception of Africa where the identical games are played but with stones or fruits, never with astralagi. This is because these particular bones have a more serious role to play in the life of the people, as we shall see later.

Perhaps the most universal of leisure activities indulged in by man is the taking of narcotic drugs. A considerable range of substances is employed, different ones being accepted in different societies. Of the socially accepted drugs in our own community tobacco in its various guises is perhaps the most widely used.

Pastimes and toys: (a) shoulder blade of sheep for making counters; (b, c) Eskimo pin and tube catching games; (d) Eskimo gambling counters; (e) Egyptian dice; (f) Roman bone doll; (g) chess-men—knight, rook (warrior) and king—11th century, Scotland; (h) Eskimo counters; (i) Eskimo pecking bird toy; (j) draughtsman, 11th century, Scotland.

67

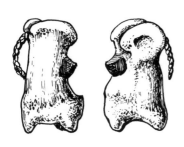

Dolls, Matto Grosso, Brazil.

Traditionally this is smoked in pipes, and bone would seem to be a singularly inappropriate material for these; however, the Lapps and Eskimos had little alternative. As with any pipe, once a carbon layer is deposited in the bowl even a bone pipe would not have been objectionable. Tobacco was tamped down with bone or ivory 'pipe stoppers' in briar and clay pipes.

As well as being smoked, tobacco can be taken in the form of snuff by sniffing it up the nostrils or by placing in the lower lip. For the snuff-taker a bone snuff-box and a bone snuff-spoon are valued possessions. The Eskimos use snuff tubes—a heroic method of snuff application. For many of the other drugs commonly used in other parts of the world there are special bone spatulas and crushing devices. The only objects we have included in the illustration are opium or lime boxes. Part of London's dockland was notorious for this drug, hence the name Limehouse.

Bones are pressed into service to beautify the human body—to improve upon nature. In this field one man's beauty is another's ugliness for tastes in such matters vary enormously. Bone pins piercing the nose are much appreciated in certain quarters, and the wearing of lip studs and bone lip spikes supposedly enhances the attractiveness of young women in some primitive cultures. Enlarging ear lobes by wearing large bone ear plugs also has its appeal, and in some communities, including our own, tattooing is popular, that in Samoa and among the Maoris being particularly striking. Indeed, the patterned scoring of the skin is common in many societies and once again bone instruments are employed. A restrictive use of bone was in the construction of chastity belts for girls in the 19th century; these had a grating of bone (non-rust) to prevent the child from masturbating.

There are more ephemeral methods of improving upon nature, as witness the cosmetics industry. Current fashion emphasises the eyes and the black liner that is used has a long pedigree. The Cleopatra look from ancient Egypt was produced by the black pigment charcoal—kohl—kept in bone tubes and applied to the eyes with a bone kohl-stick. Ancient Egyptian toilet trays, too, were made of bone and used as palettes just as they are today.

The other important item in the armoury of appeal is the 'hairy diadem which on them doth grow'. As well as frequent grooming, hair has to be kept in place, and pins and combs accomplish this. Naturally, decorated combs and pins add to glamour and are worn by both sexes in several socieities. Again as expected, comparable hair ornaments are found in every continent.

A further way of decorating one's person is simply to wear such ornaments as bone rings and bone bangles, the former for fingers

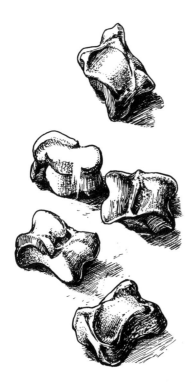

'Bones'—astralagi.

68

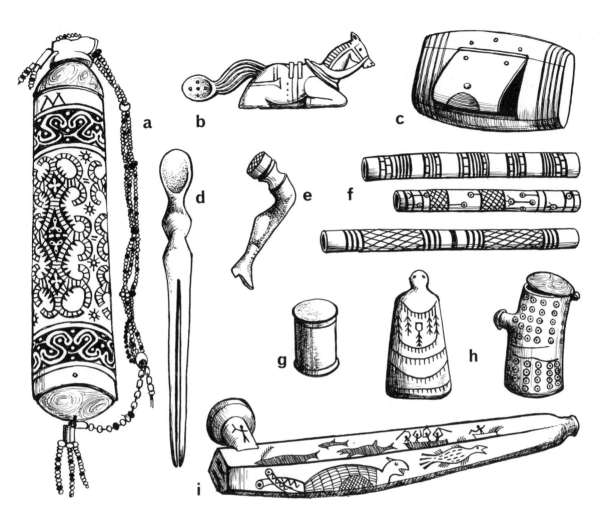

and the latter for legs or arms. By far the most decorative objects worn are necklaces, necklets and bracelets, in which bones and teeth figure prominently. Sometimes, as in the Ecuador necklace illustrated, only teeth are used, in other cases just bones, such as the claw bones from South Africa. The necklet from the Solomon Islands is made from alternating fruit-bat canine teeth and fish vertebrae. False teeth have been made of ivory or bone from time immemorial, the enamel of hippopotamus incisors, by far the strongest and whitest, being the most favoured.

In most cases we accept necklaces and other objects which are worn as being pure decoration. It may well be that such notions were entirely alien to the people concerned. Just as women in North Africa wear necklaces of coins, their decoration being also their wealth, so too can bone and tooth necklaces represent currency.

Narcotics: (a) opium tube, Indonesia; (b) snuff-spoon, Scotland; (c) snuff-box, England; (d) snuff-spoon and comb, South Africa; (e) pipe-stopper, England; (f) Eskimo snuff-tubes; (g) opium-box, China; (h) Eskimo snuff-boxes; (i) Eskimo pipe.

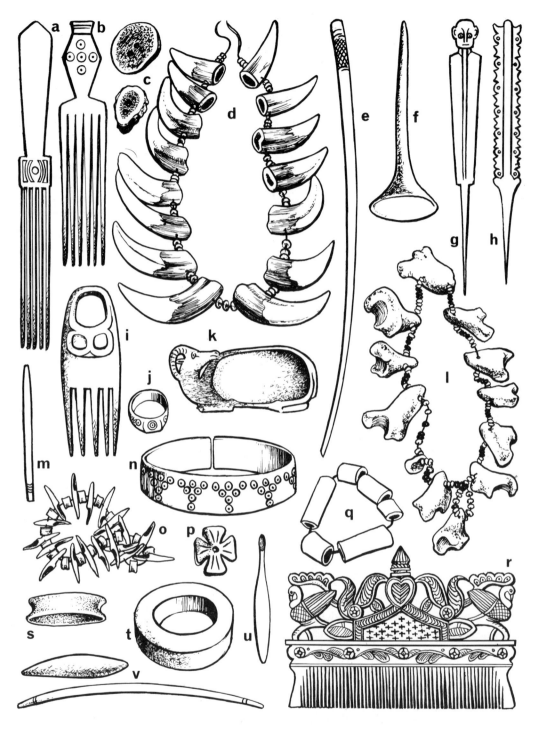

Ornaments: (a) comb, South Africa; (b) comb, West Africa; (c) buttons of red-deer antler, Scotland; (d) necklet of peccary teeth, Ecuador; (e) hairpin, Central Africa; (f) hairpin, Congo; (g, h) hairpins, Malaya; (i) Eskimo comb; (j) ring, Tanzania; (k) toilet tray, Ancient Egypt; (l) necklet of terminal phalanges, South Africa; (m) kohl stick, ancient Egypt; (n) bangle, India; (o) bracelet of canine teeth of fruit bat and fish vertebrae, Solomon Islands; (p) Roman button; (q) bracelet of bird wing bones, New Mexico; (r) comb, North West India; (s) lip stud and (t) woman's ear plug, India; (u) English ear scoop; (v) nose pieces, New Guinea.

On Manam Island in New Guinea, the women wear necklaces of dogs' canine teeth; five canine teeth are equal to one currency unit, the equivalent of a shilling (5p). In the Solomon Islands necklaces of dolphin teeth are worn; these are worth ten a shilling, and it requires an astronomical number to purchase a wife. In Fiji strings of lower jaws of fruit-bats are used as currency, and sperm whale teeth or tambua are paid to the government as tribute. Among the Shoshone and Bannock tribes of Idaho and Montana the canine teeth of the wapati (elk) are used as currency, one canine tooth equals 25 cents (10p). They are only used for transactions among themselves and never in dealing with the palefaces, and these same teeth are used as ornaments for dresses.

Although the ornaments displayed may be a measure of the wealth and standing of the individual concerned, there can be further significance in the type of ornament. For example, in parts of Nigeria a necklace of snake vertebrae is worn as a protection against snake bites, and on other occasions, as with the Hausa of West Africa, a string of bones may be simply prayer beads.

In most parts of the world bones are believed to possess powerful magical properties. In the Naga Hills of India the jaw bones of a dog hung from the lintel of the door will keep evil spirits at bay, for when such spirits approach the house, the dog spirit will bark and frighten them away. A suitably inscribed sheep scapula will serve as a protection against scorpions in Algeria, and in fact throughout North Africa the scapula of a sheep is hung up for luck. In Devon, half the pelvic girdle of a dog will protect one from the evil eye; the cannon bone of a sheep will protect from rheumatism in Huntingdonshire and the astralagus of a sheep will ward it off in Suffolk. A mole's foot will serve as a charm against cramp in Hampshire, a wood pigeon's also in Hampshire and a moorhen's in Northamptonshire. And in Scotland a hare's foot was worn in the undergarments of pregnant lassies to prevent miscarriage, by warding off evil spirits.

To ensure success in fishing, bone models of fish used to be (and may still be) carried by the fishermen of Great Yarmouth, Norfolk; indeed they would never venture forth to fish without them.

As well as having all these powers, bones also predict the future and in Africa, astralagi are used in divining. In northern Nigeria divining chains, or agbendi, consist of four pieces of sculptured turtle carapace; when the chain is thrown on the ground the combination of smooth and sculptured plates appearing uppermost will provide the required information. Southern Nigerians use crocodile scutes in the same way.

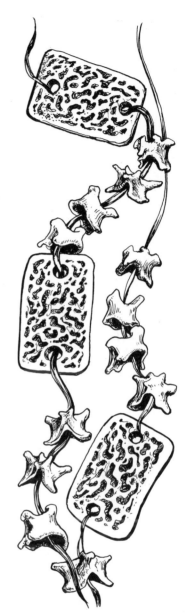

Nigerian divining bones—snake vertebrae and turtle bones.

Certain bones and teeth, if ground up and swallowed, are supposed to cure particular ailments. In China dragon's teeth are especially potent—the dragons in question are rhinoceros, orangutang, panda, pig, water buffalo and *Hipparion*, an extinct three-toed horse. Last year one of us spent half a day in a market in Kampala, Uganda, trying to bargain for a piece of arthritic buffalo bone; as it was 'powerful medicine' it cost us 10p.

Haida Indian totem figures engraved on whale vertebrae.

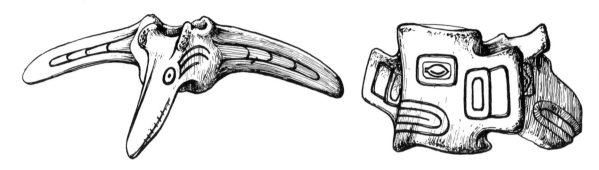

All the instances of bone charms and medicine are obviously aimed at the improvement of man's sorry lot but this is, sad to say, not the complete story. They can also be used malevolently. In southern Australia the pointing bone can cause a person's death by magic. Goya in his drawing 'A caza de dientes (Out hunting for teeth)', in his series *Los Caprichos* published in 1799, commented: 'the teeth of a hanged man are very efficacious for sorceries; without this ingredient there is not much you can do. What a pity the common people should believe such nonsense'.

The Haida from the north-west coast of America used whale vertebrae to fashion totems. These were only slightly carved, the natural projections of the bone being transformed into figures; the spine became a beak, the lateral processes wings, and the body of the vertebra a face.

Masks used by Eskimos in their ritual dances were sometimes shaped out of the light spongy bones of large whale vertebrae.

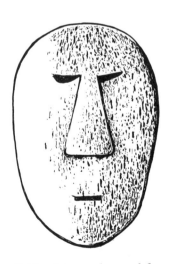

Eskimo bone mask carved from a whale vertebra.

In 385 A.D., during a conflict with the authorities, St. Ambrose barricaded himself in Milan cathedral. During the siege he dug up two Palaeolithic skeletons from under the church floor. It was 'revealed' to him that they belonged to Gervaise and Protasus, who had suffered martyrdom during Nero's reign. These bones drove demons from lunatics, restored sight to the blind and protected Ambrose from the soldiers—'proof' (if such were needed) that they were genuine relics of Saints.

Bones of mammoths and woolly rhinoceroses were resurrected into dragons for worthy knights to defeat in mortal combat. It is not at all unreasonable that some of the woolly rhino's human contemporaries should have had the opportunity to join the forces of righteousness.

The making of bone china, glue and fertiliser are all uses to which bone is currently put, but we cannot think of any more unusual use of bone than that of creating Christian Saints out of them.

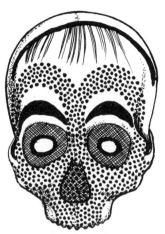

Skull of chief decorated with tattoo pattern worn in life, Naga Hills, India.

Skulls of dogs sacrificed to Ogun, God of Iron, in Old Idanre, Western State, Nigeria.

7 Bone culture

Among the peoples of the Arctic, there developed a true bone culture. Bone and walrus ivory were to all intents and purposes the only suitable materials available for manufacturing the products necessary for survival. Life in the Arctic has always been hazardous for man, yet the Eskimos have more or less flourished there for innumerable generations. Only now is their end in sight with the opening up of the Arctic for its oil and mineral wealth. Delicately balanced societies are meeting their demise as a result of their encounter with the affluence of 'civilised' society. In time an Alan Moorehead will write of 'The Fatal Impact' on the Eskimo. Already we have the situation in which walrus ivory is sold to Japan and the ivory carvers of Yokohama turn out polar bears, walruses, seals and 'eskimos' for export to Alaska where they are sold to tourists as 'native' work. The Canadian government is actively encouraging Eskimo art, but sad to say that art does not any longer have the vitality of the objects previously made for use in the business of living.

Cord attacher.

But the Eskimo way of life has not yet entirely vanished. Drs Franz Boas and E. W. Nelson of the Smithsonian Institution recorded a vast amount of information during the latter part of the 19th century, and more recently Professor Robert Gessain, Director of the Musée de l'Homme in Paris, has written extensively on the Eskimos of East Greenland.

Although we have already discussed many of the uses which Eskimos have made of bone, in this chapter we are concerned with how the Eskimo manages to survive in one of the most hostile environments on earth. Bone has played a vital part in allowing him to do so.

Hunting and fishing take place outside the winter season, each season demanding different skills. In the spring, hunting is on land and ice. When the seals come out to sun themselves on the ice, the seal hunter, dressed as a seal and simulating its movements, crawls towards his prey. If the seal can be cut off from escape, it is a simple matter to club it to death.

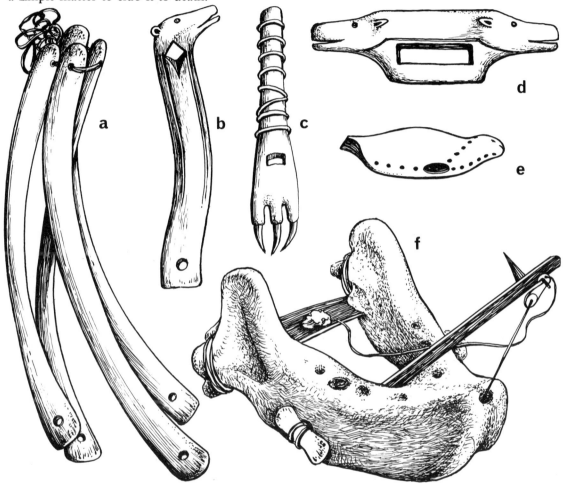

The downfall of the seal lies in its insatiable curiosity. A particular hunting trick has been learnt from the polar bear which draws seals by scratching the ice. The seals, overcome by curiosity, emerge and are slain and devoured. The Eskimo fashions imitation ivory or bone bear paws bearing three claws which are then used to attract seals, ready for harpooning. When seal pupping is in progress, usually in excavations some 5 to 10 feet deep in the snow, the Eskimo digs into the cavern, whereupon the alerted mother escapes and the pup is dragged out with a special crook at the end of a pole and is killed by firmly treading on its chest.

Hunting and trapping equipment: (a) seal club; (b) arrow straightener; (c) ice scratcher; (d) drag handle; (e) qanging; (f) trap made from lower jaw of walrus.

Alternatively, it may be secured by a thong through the back flipper and thrown back. As Dr Boas described, 'it dives at once, crying pitifully. The dam returns to her young and attempts to draw it away. When she is seen a harpoon is plunged into her body'.

The carcass is dragged across the snow by means of a kind of toggle, termed a 'qanging', attached by a thong to a drag handle. The thong is pushed through the seal's lower jaw, and the qanging serves as a stopper.

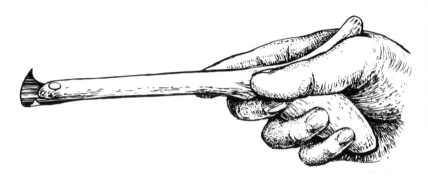

Plane in use.

Caribou and other deer are hunted by arousing their curiosity. For this exercise two Eskimos work together. When the animals see them, the men walk away from the deer which proceed to follow; when cover is reached, one man stays behind while the other continues walking. On coming level with the hidden Eskimo, the deer receives a spear in its flank, which cripples it and thus allows it to be despatched with the minimum of effort.

Small game is caught by means of traps built on the mouse-trap principle. A spike-bearing arm is held back by a small bone catch which is released once the bait is touched; the spring is a twisted sinew tightened by two bones. The specimen we have illustrated is made from the lower jaw of a walrus. Other small game may be hunted with ivory bolas or with bow and arrow. The bows may be made of deer antlers. For both arrows and spears, the shafts are straightened over a fire using a bone shaft-straightener.

Bone knives are used to skin the carcases, and the pelts are pegged out with bone pegs and scraped with bone tools to remove the flesh and fat, bone polishers being used for softening the hides. Small skins are soaked in urine to remove the fat and are then stretched and worked by hand. Special bone combs are used to prepare deerskin.

For making thongs there are sinew shredders as well as sinew twisters. Thereafter bone needles and thimbles are used for

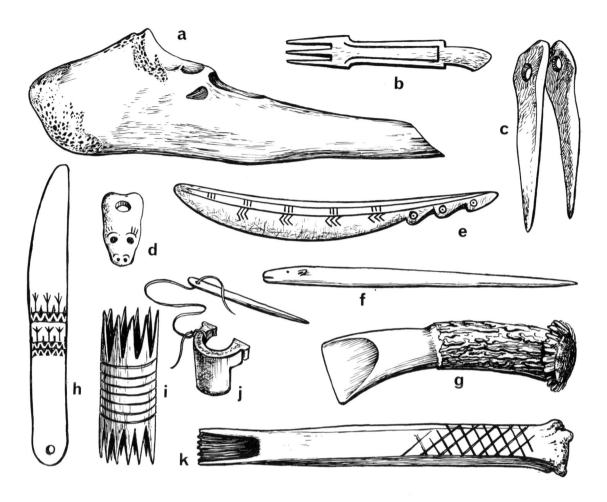

sewing and finally the garments are fastened with bone or ivory buttons and toggles.

Travelling over the snow is accomplished by dog sleigh with runners made of walrus ivory. Bone mallets are used for knocking off ice from the runners and snow tappers for removing snow from clothing. When travelling over frozen snow or sea-ice, ivory ice-creepers are tied to the soles of the shoes. Because of the snow and bright light, snow goggles are worn to protect the eyes. Most examples we have seen in museums seem to have been made of wood but the one lent to us by Arthur Bourne is fashioned out of a caribou long bone.

With the coming of winter, snow-houses or igloos are built. Blocks of snow are cut from snow banks formed during a single storm, for if the blocks contain several layers of snow they break. Ivory or bone snow-knives are used for cutting the blocks as well as bone shovels.

Tools used in preparation of pelts: (a) skinning knife; (b) sinew shredder; (c) pelt pegs; (d) toggle; (e) snow knife for cutting blocks of snow for igloos (nothing to do with skins); (f) sinew twister; (g) scraper; (h) knife; (i) deer-skin cleaning comb; (j) needle and thimble; (k) flesher.

Snow creepers.

In addition to land-based hunting, fishing is also carried on, the most familiar type being through a hole in the ice. An ice-pick is used to make the hole and bone scoops are used to remove the pieces of broken ice. A line is lowered with a sinker or ivory fish bait and when the fish come to investigate they are speared with pronged fish-spears. In other cases, as the fish gather round the bait or ivory sinker they are caught on hooks pulled up from below. It is worth noting that the fishing lines are not baited with anything edible but merely with pieces of bone or ivory which are enough to attract the fish. Indeed, salmon appear to be taken by polar bear teeth. During the summer, fishing with nets is carried on. Different types of fish-trap are made employing floats of wood and ivory sinkers. Again the instruments used for making the nets are all made of bone or ivory, such as netting needles and net spacers, thread reels and shuttles, the last two generally functioning in either capacity.

Carved bone sticks or fish stringers are used for carrying the fish; the instrument is thrust through the gills and out through the mouth, so that several can be threaded together.

Eskimo wearing his snow goggles made of caribou bone.

During the summer most hunting is from boats in open water, and whales are hunted from large, open boats or umiaks—a most hazardous undertaking. Umiaks were normally used for transport and were associated with women whereas the male hunting vehicle was the kayak which only carried a single man. The kayak has a light wooden framework completely covered by skins apart from the central orifice for the man. The paddle has the working ends made of bone and the stern and prow of the kayak have ivory points in the shape of a penis emphasising the masculinity of the craft. There are thirteen cross-straps running around the kayak which serve to secure the hunter's equipment. This consists of a harpoon (the main weapon), two lances, a bird spear and a short, stabbing spear. The harpoon has a wooden shaft with a blunt ivory base which is used for cutting ice and hummocks impeding the progress of the kayak. There are two ivory side projections which insert into holes in a decorated throwing board. The head of the harpoon shaft has an ivory cap into which an ivory spike is fitted, itself fixed to the shaft by a thong. The head, inserted on the spike just before the harpoon is hurled, is attached by a long cord to a sealskin float.

The Eskimo hunter has to serve a long apprenticeship before he can master the kayak. It is an exceedingly efficient machine, so much so that virtually every movement in the hunt is determined by the positioning of the armoury. The gradual evolution of the kayak has resulted in the present highly stylised version. From the front end the first cross-strap secures the points of the bird spear (and a simple spear for baby seals). After the Y-shaped cord the third and fourth straps hold the white cloth camouflage sheet which enables the hunter to approach within striking distance of his prey, the seal simply imagining he is just another piece of floating ice. Also held in the double cross-straps are instruments for cleaning skins, extracting the harpoon head from the corpse, the short stabbing spear for piercing the heart, and the anterior supports of the harpoon line holder. The double straps 5 and 6 contain bone knives and snow and ice scrapers for cleaning off snow and ice from the surface of the boat. Straps 7 and 8 hold the foot of the main support of the tripod on which is placed the harpoon line. To the right, a short strap, no. 9, holds the support for the harpoon shaft which with its throwing board points backwards. Cross-strap 10 secures the bases of the two lances. The double straps 11 and 12 hold the two supports of the float which is placed slightly towards the left. The exact strength by which the float is held requires considerable skill, as it is vital that it should be securely held in normal progress but should slip off

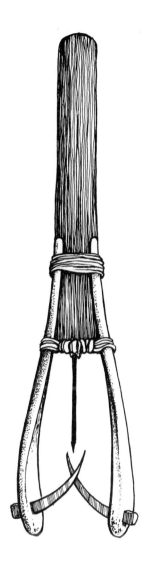

Head of fish spear.

79

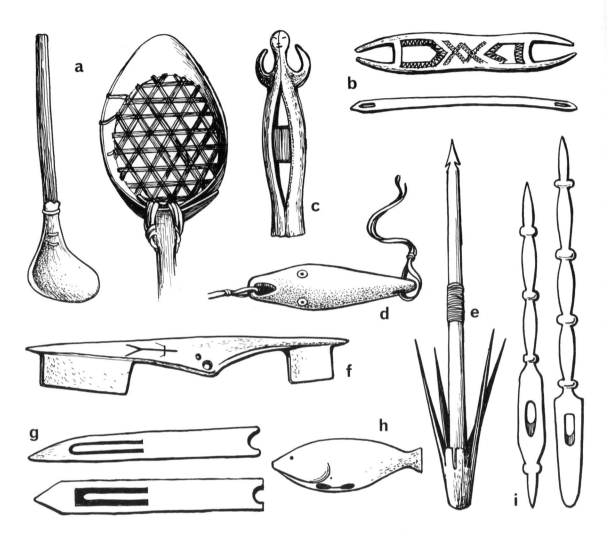

*Fishing equipment: (a) ice
scoops; (b) net shuttles; (c)
thread reel; (d) sinker; (e)
fish catcher; (f) net spacer;
(g) netting needles; (h) bait;
(i) fish stringers.*

rapidly once the harpoon strikes. Young inexperienced hunters
are frequently killed by drowning because the float is held and the
prey drags the kayak. Finally the thirteenth cross-strap secures the
tips of the lances.

Immediately in front of the hunter is the harpoon line carefully
wound with the harpoon head in the centre. The rim containing
the line is made of bone as are the three supports. The cross-straps
are tightened or loosened by the ivories. The main support is
decorated with ivory seal emblems. In the drawing, a support for
the harpoon shaft is part of the right leg of the tripod. The small
stabbing spear is present, together with a string of bone suture
needles, used to close the wounds of prey once it has been killed.
To the immediate right of the main central leg of the tripod are
two bone tools.

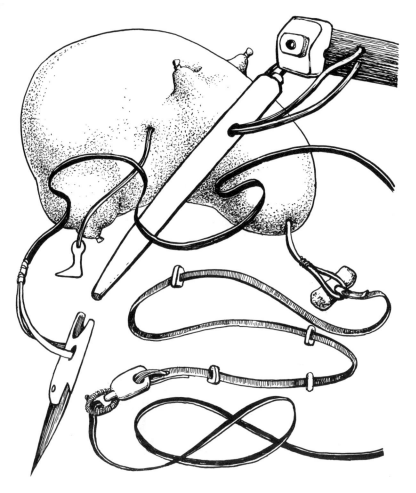

Harpoon and float: harpoon head attached by long thong line to strap, which in turn is attached to the float—an entire baby seal inflated. The harpoon head is slotted on an ivory spike, which is hinged on the main shaft of the harpoon.

Harpoon head.

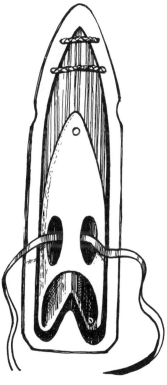

The harpoon line runs to the right of the hunter to attach to the float. To make the float a young spring seal is caught, cut around the muzzle and anus, as well as the flippers just behind the claws. With a knife the skin and fat are separated, then the viscera, flesh and bones. The carcass is then held by two men and a third empties the contents through the mouth. It is handed over to the women who turn it inside out and clean it; thereafter it is hung on the ceiling above the lamp for several days, sewn up and has a plug inserted for inflating it. Two lines are run through it so that it can be secured to the cross-straps of the kayak.

Large seals are also emptied of their contents but with a wide transverse slit across the chest. The skins make excellent cosy sleeping bags.

When a seal is sighted within range, the harpoon is armed by inserting the head on the ivory spike. The harpoon is gripped in

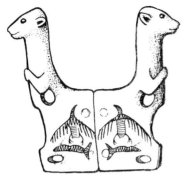

Harpoon rest.

the right hand with index finger and thumb at the appropriate notches of the throwing board. The arm is first lifted to rotate the harpoon so that it now faces forwards then drawn back; as the throw begins only the throwing board is held, giving extra leverage for the throw. When the seal is struck, the harpoon head penetrates the body, the ivory spike separates from the head and the shaft falls back at the hinge with the spike. The harpoon shaft and spike float on the water to be retrieved by the hunter. The head, attached by up to 30 yards of line to the float, is turned round in

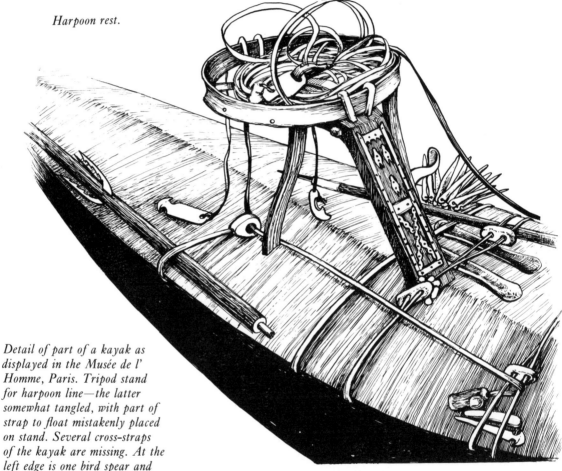

Detail of part of a kayak as displayed in the Musée de l' Homme, Paris. Tripod stand for harpoon line—the latter somewhat tangled, with part of strap to float mistakenly placed on stand. Several cross-straps of the kayak are missing. At the left edge is one bird spear and on the right a short stabbing spear for piercing seal hearts and a bundle of wound-closing suture pins. Two skin-preparing tools are shown under the cross straps. At the bottom right of the drawing are two further wound-closing tools.

the flesh once the animal dives and the line is pulled taut by the float. The float impedes the animal's movements and marks the position where it will surface to breathe. When it does so, the hunter is waiting and the animal is stabbed through the heart with the short spear. A serrated bone scraper is used to extract the head and push back any viscera escaping from the wounds, and

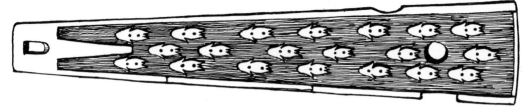

Throwing board.

holes are closed by bone suture pins. The body is inflated by blowing into a hole made in the neck, and the front flippers are tied together with thongs and bone toggles. The carcass is then towed back, secured by a hook, or in the case of large animals, with an ivory spike through the nose.

When large animals such as walruses are hunted, several hunters work together. This type of hunting involves close-in work. The lances have detachable heads and as a lance is thrust into the body the head detaches. Immediately a new one is attached and a further thrust made. After the kill, when the animal is cut up, the hunters retrieve their own lance heads, identifiable by their individual decorations.

As well as hunting the larger animals, the hunter also catches birds using a bird spear which has a fine metal spike at the tip and, halfway down the shaft, three barbed points. The bird spear is thrown by means of the throwing board, again over-arm, although some hunters do it under-arm. Spears are thrown with astonishing accuracy over 30 to 50 yards. The under-arm action allows the spear to skim the water to catch birds floating at the surface. The other method is to throw the spear high in the air so that it comes down and traps them. The hunter captures the birds alive, killing them by gripping the beak in his teeth and twisting the body.

In case we have left an impression that the Eskimo is particularly callous with regard to his prey we must emphasise that this is far from being the case. The Eskimo has a deep respect for the animals of the Arctic: indeed the killing of game is beset by innumerable taboos. He is allotted a certain amount for his life; if he takes too much it is a sign of his own impending death. He never kills for amusement, only to survive. For some Siberian tribes the hunt is considered exactly equivalent to murdering his fellow man.

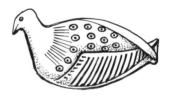

The ornaments and decoration with animal motifs of exquisite tenderness illustrate the affection and respect which the Eskimo holds for the creatures with whom he shares the Arctic wastes.

Eskimo carvings: ear ornaments, container and sinkers.

8 Bones as media

From the earliest examples of Man's creative activity bone has been the medium of this expression. We know that one of the very first steps on the road to mankind was when our ancestors became carnivores. From eating meat it is but a short step to discovering the usefulness of the bones in the remains of one's food. Thereafter it is not at all difficult to envisage the fortuitous scoring of bone — the accidental making of a simple primitive pattern. This could have been accomplished in the course of hacking off chunks of

Limb bone of a sabre-tooth tiger carved by prehistoric man in California.

meat. Perhaps the simplest deliberate marking of a bone would have been akin to that seen on the shin-bone of an American sabre-tooth tiger from the Rhancho La Brea tarpits in Los Angeles.

Bones decorated by prehistoric man.

This consists of two parallel grooves running obliquely across the main part of the shaft. From such simple carving there seems to have evolved patterns of lines crisscrossing, frequently producing a characteristic herring-bone.

This abstract decoration continues unbroken to the present day. Fortunately for our purpose, prehistoric man did not stop there but began to decorate his bone tools in other ways. A spear shaft straightener has small ponies engraved on it. In other instances the shape of the bone itself must have suggested the form of different animals which could be carved in firm relief. The crouching deer from France is an especially elegant example. The strange antler tool with the hole drilled at the working point has the handle-end transformed into a leaping horse. Similar carvings are made in other materials and there are innumerable cave paintings, rock engravings and even clay models, but even at this time bone was obviously considered an appropriate medium for such work.

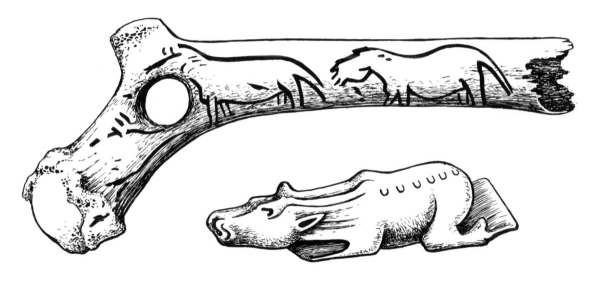

When it came to the portrayal of man himself and more especially woman, bone and ivory were frequently employed. There seems to be a deeply ingrained belief that prehistoric man only portrayed female torsos paying no attention to details of faces, merely to accentuated breasts, bellies and buttocks. Indeed, the beginning of every book on the history of art illustrates this point by showing the Venus of Willendorf. This helps to perpetuate the myth that either prehistoric woman was pretty unattractive or prehistoric man's tastes were somewhat unsophisticated, unlike nowadays. All this is very much a travesty of the real situation. Rosemary

Shaft-straightener engraved with horses and carved crouched deer, both made by prehistoric man some 20,000 years ago.

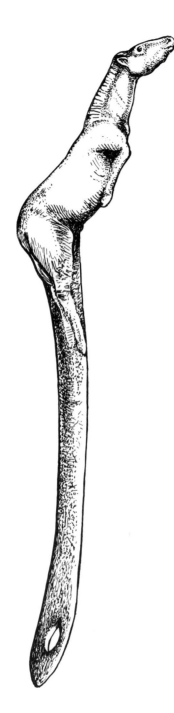

Awl made from antler with leaping horse handle.

Powers of the Natural History Museum has assembled a dossier of all the prehistoric portrayals of the human form. From this monographic compilation it clearly emerges that the variety of womanhood was exactly as it is today. Women of all shapes and sizes were delineated, just as attractive or otherwise as nowadays.

As well as the naturalistic Venuses, there were more symbolic, stylised figures. Rosemary Powers' study shows that men were also portrayed, and furthermore that their faces were carved and painted. The carved heads reveal a skill and subtlety that has rarely been equalled. Perhaps one of the most beautiful examples of Palaeolithic art is the tiny ivory head of a girl from Brassempouy in France, carved in mammoth ivory and showing the most delicate modelling. Another beautifully carved head comes from Dolni Vestonice in Czechoslovakia. This one is noteworthy by virtue of its gender—it is clearly that of a finely-featured man. The drawing is of a cast of the original, kindly given to us by Rosemary Powers.

The highly stylised pear-shaped female torsos seem to have been worn as ornaments. Although the main emphasis seems to have been on the female of the species, bones were often carved to represent phalli: some had phalli engraved upon them and some antlers were transformed into bifurcated phalli. One can only surmise the purpose of such representations. Perhaps they were the counterparts of the vulvoliths or stone vulvae which were also frequently portrayed by prehistoric man, and indeed his modern descendants.

Later examples of bone carvings in which the form of the bone has determined the alignment of the human form are well shown in some Egyptian work. A kohl tube, that is a container for eye make-up, as mentioned earlier in this book, is constructed out of the hollow long bone of a deer or goat. On the thick outer part has been carved a female form in relief. The form of the bone virtually suggested its use: the embellishment of the person for whom it was intended. This example comes from the XIXth Dynasty of ancient Egypt. The other specimens are 4th or 5th century A.D. from the time when Christianity flourished in North Africa. Today the Coptic Christians survive only in Ethiopia.

The three examples illustrated show how the fragments of long bones have been utilised to best advantage: the two upright figures with their heads appropriately turned, the young lady with hands folded and attired in flowing dress, and the male nude with the bone broken in manner most discreet. The Nereid, or water-goddess, in contrast is aligned prone along the length of the bone in a frankly seductive pose.

In medieval Europe there seems to be little of note in the bone or ivory world. Art in all its manifestations was dominated by matters ecclesiastical, and in Germany and central Europe, and to a lesser extent England, the most elaborate illustrated panels of religious events are delineated. These intricate carvings in ivory are not to our personal taste but we have illustrated one of the least elaborate we could find. This is a representation of the Holy Trinity, showing the Father, the Son and the Holy Ghost. At least it demonstrates that, in medieval times, artists had no hesitation in portraying the Holy Ghost. God the Father stands with His Son in crucifixion posture athwart Him, with the Holy Ghost in the shape of a dove issuing from the mouth of the Father to that of the Son. Since that time Father and dove seem to have fallen into oblivion.

A more secular use of teeth and bone is in scrimshaw which developed during the hey-day of the whaling industry during the eighteenth and nineteenth centuries. The sperm whale has teeth only in its lower jaw and these are deeply embedded in the gums. After a sperm whale was caught the lower jaw was either exposed on deck or towed behind the ship for about a month. By this time tooth extraction presented no difficulty. The second or third mate was in charge of the distribution of the teeth among the sailors. When fresh the teeth were easy to work; the main outlines of the picture were made with a jack-knife and the fine lines with a pickwick, i.e. a sailcloth needle.

Ivory head of girl, from Brassempouy, France and man from Vestonice, Czechoslovakia.

After being etched, the teeth were then buffed and polished with damp wood ash. The drawings were coloured with lamp black or tobacco juice which was rubbed into the scratches. After drying, a final polish was given.

T. K. Penniman, of the Pitt Rivers Museum in Oxford, noted (in 1952) that the seamen carve on these teeth 'more or less beautiful designs, or engrave pictures according to their abilities, some designs being pleasing to persons of sensibility, and others resembling the cruder designs tattooed on their bodies'. Not surprisingly, ships figure prominently but so too do a number of elegant ladies.

As well as the teeth, the thin jaw bone or pan was also used for scrimshaw work. Ships and whaling scenes were the normal motif. From pan were fashioned 'longing sticks' on which were depicted such evocative pictures as bowls of fruit, and home!

Throughout history bone has been a recognised medium but it took its place with many others that were constantly to hand. For a brief period, as far as we are aware, it was, to all intents and purposes, the only one available. During the Napoleonic wars the

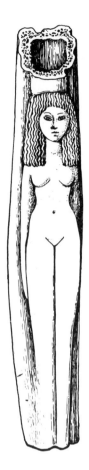

French prisoners of war, incarcerated in British prisons, produced a wealth of bone carvings that have never been rivalled before or since. These items are now very highly valued and are much sought after by collectors.

Initially, the prisoners were housed in prison hulks anchored off the coast, in Plymouth Sound. The overcrowded conditions in these hulks must have been pretty appalling. Most of the P.O.W.s were later housed in a specially built prison camp at Norman Cross, near Peterborough, from 1797 to 1814. Here 7500 prisoners were kept, of whom 1770 died in captivity, about a 1000 of them during an epidemic in 1801. Sheila Selby has recently made a model of the camp as it then was: there were sixteen prison huts 100 feet long by 22 feet wide, two storeys high, each housing some 500 men. Two of these huts served as a hospital, and each group of four huts had access to a quadrangle about 300 feet square where the prisoners could pass their time away.

Bone kohl tube from ancient Egypt.

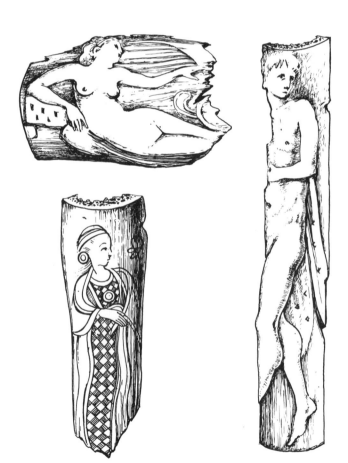

Human figures engraved and carved out of bone, 4–5th century Egypt.

88

In the Peterborough Museum there is an inventory for the prisoners' rations and the notable feature of it is that it does not vary from day to day nor from month to month. A typical day's provisions were as follows:

Bread	Beef	Beer	Peas	Potatoes	Salt	Cheese	Butter
6600 lb	2050 lb	4000 qt	500 lb	3070 lb	85 lb	5101 lb	385 lb

To say the least of it, this was a fairly monotonous diet, albeit adequate to sustain life. There is no indication of any ill-treatment, only an excruciating boredom. The only relief, the only recreational activity possible, was carving. The most readily available material which these captive soldiers and sailors could work were bones from the cook-house. Hence, for twenty years the sole gainful occupation was carving beef bone.

In fact some wooden ship models were made including a few constructed out of fine wood shavings and some elaborate pictures

Byzantine casket plaque of a griffin, 12th century.

Holy Trinity, early 14th century, England

89

Ivory medallion made for Englishman by Italian in Rome, 1724.

Scrimshaw engraving on tooth of sperm whale.

were made of straw. Without doubt, bone work was considered the most important—so much so that the prisoners melted down their ear and wedding rings to manufacture the tiny pins for fixing the pieces of bone together.

Although prison security was tight and no prisoners were allowed out on parole, they were permitted to sell their carvings at the gate each day, albeit under the eyes of a watchful and heavily armed guard. These prison markets became justly famous and the gentry would come from far afield to buy the carvings from the prisoners to whom local girls in turn would sell fruit. Dr T. Walker was the author of a classic work: *The Depot for Prisoners at Norman Cross, Huntingdonshire, 1796–1816,* published in 1913 when he was 80 years old.

In his speech proposing a toast prior to the unveiling of the French Prisoners' Memorial on 28th July 1914, he commented that 'they for eleven years had been shut out of female society, and had never spoken to a woman unless it happened to be their turn to go to market and they had said "Ma Cherie" to an English girl in an unknown tongue—though he dared to say there was a language which was known everywhere'. Indeed, it is interesting to record that after being repatriated to France one prisoner returned to marry a local girl and settled down; his descendants live in the same area to this day.

Going to the gate to sell carvings was the only possible contact with the outside world, hence the prodigious productivity in bone work. Some of the more elaborate productions must have been able to command high prices and it is stated that many prisoners returned to France after the war with as much as £1000, which in the early 1800s was a considerable fortune.

Dartmoor boasted a similar prison which was a more substantial edifice than the wooden barracks of Norman Cross, and indeed still houses prisoners. Built originally for French prisoners, Dartmoor was begun in 1806 and completed in 1809, when 2500 Frenchmen were marched over the Moor from Plymouth. When the United States declared war on Britain in 1812, 250 Americans were marched through the snow to the prison. There is no record of bone work by Americans; their only recorded contribution to prison culture would seem to be the following lines written by Charles Andrews:

Any man sent to Dartmoor might have exclaimed
Hail horrors! Hail thou profoundest hell!
Receive they thy new possessor.
For everyone ordered to this Prison counted himself lost!

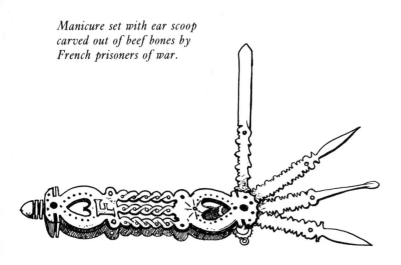

Manicure set with ear scoop carved out of beef bones by French prisoners of war.

In spite of this, the industrious Frenchmen settled down to emulate their comrades at Norman Cross and carve the bones from the cookhouse.

The remarkable feature of all the bone carvings of these prisoners is that they clearly reflect an all-male society. There are few items of a purely domestic nature and few one would associate with the vanity of the fair sex. A possible exception is the ornate filigree-type manicure set with earscoop. The basic products carved by soldiers and sailors were gambling devices. It seems a reasonable surmise that this bone culture began as a means of creating the wherewithal for games of chance to help pass their days of enforced inactivity—hence dice, dominoes, cribbage, cards, and small bone boxes constructed to house the games.

There are usually three dice in these boxes, which puzzled us until we noticed that in several Parisian cafés the local men were playing *quatre vingt et un*, presumably a survival from Napoleonic times! Presumably such gaming sets would have attracted the interest of the guards who might have either bought or bartered for them. Thus may have begun an escalating industry, with ever more elaborate pieces being developed until the prisoners' market became established.

This male-orientated industry produced innumerable tobacco tampers—the unusually shaped one we illustrate was made for Captain T. Lincolne Barker who was the Governor of Norman

Longing stick, scrimshaw engraving on pan—jaw bone of sperm whale.

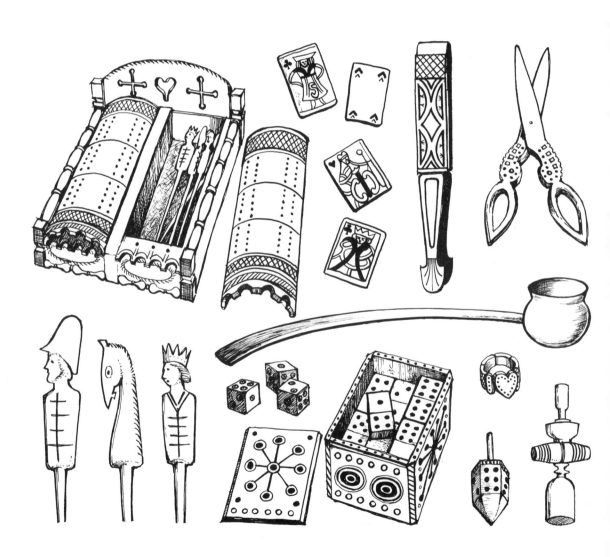

French prisoner of war work in beef bone. Games, initially for their own use: cribbage board and pegs, playing cards and dominoes, and 'quatre-vingt et un' (played with three dice). Other items of a more domestic nature include tongs, scissors, punch-ladle and tobacco-tamper.

Cross. Snuff boxes, both simple and decorated, were favourite commodities. Simple rings, dice shakers and small punch-ladles were all most likely to have been made in the first instance for local use inside the prison by both captives and captors.

As the workmanship became more ambitious and elaborate, in many ways heralding the intricacies of the later Victorian era, so an outside market gradually opened up. Needless to say, this activity was encouraged by the prison authorities.

A whole variety of miscellaneous objects were carved, many of which would seem to have been merely novelties, such as the pair of scissors. Others must have been reasonably functional such as sugar tongs, apple scoops and screw nutcrackers. Folding paper-knives, combs, back scratchers and hinged salad-forks and spoons must also have been capable of some use.

The miniature musket made for Captain Barker in 1810, as well as models of bellows, shovel, poker and tongs, tiny knives, forks and spoons, toy tables and chairs are likely to have been made for the children of the guards or for the girl fruit-sellers. Other comparatively simple works were chess sets and small crosses.

The next development was the carving of small figures. A number of crucifixes are known with Christ on the cross with two standing women at either side: presumably Mary, mother of Jesus, and Mary Magdalene. The more frequent figures, however, are soldiers, officers and ladies in bonnets. We have illustrated two from a somewhat complicated tableau: a hunter in top hat with dog in attendance together with a woman in boots, pleated skirt and low-cut blouse of markedly contemporary mien.

Napoleonic prisoner of war carvings: small figures and mechanical toy of blacksmith, cooper and drummer.

Carved figures for which there must have been an almost insatiable market were those that moved. A comparatively simple one with three men in a row was set in motion by turning the handle at the side. By means of a series of cogged wheels, the blacksmith at the left strikes his anvil with his hammer, the soldier at the right beats his drum and the cooper in the centre hammers the barrel as it turns round. Other set pieces show soldiers honing their swords on a sharpening stone. But by far the most popular were the spinning jennies with long-gowned and high-bonnetted ladies. All the workings of these jennies were open to view and some of the working models were exceedingly involved.

Spinning jenny.

One example in the Peterborough Museum, shows a lady turning the spinning wheel, a soldier and lady waltzing, a mother tossing her baby, a child moving forwards and a woman making tea. The more elaborate of these mechanical toys seem to us to lose much of their charm, although one cannot deny the incredible intricacies of the workmanship. In the more extreme and involved examples gimmickry seems to have finally carried the day. Perhaps this is a reflection of a 'keeping-up-with-the-Jones' syndrome among the landed gentry of Huntingdonshire, newer customers demanding ever more complicated working models.

Many of the objects constructed by these prisoners are so intricate that it is extremely difficult to believe that they are made of bone. Small 'filigree' baskets and the most incredibly ornate cases for clocks were constructed with the utmost delicacy, looking more like tracery of lace than something out of beef bone. This same complex tracery is used to decorate bone boxes of various shapes and sizes—sewing boxes, ditty boxes and the like.

Small bone houses were made with doors and windows, verandas with tables and chairs and even aviaries complete with caged bone birds—a.sad commentary. Again we have illustrated the smallest example. By far the most elaborate is a huge chateau, four storeys high with two clock towers. There is a water-wheel, and in fact running water drives this wheel which in turn moves all the many figures on, in and around the chateau. Two men saw a plank, a couple promenade in the garden and a whole variety of domestic tasks are performed by innumerable servants. This elaborate edifice simply does not give the impression of having been painstakingly constructed out of tiny pieces of bone, even down to the fences.

Perhaps the most dramatic of the working models that must have been an absolute hit with the English gentry were the guillotines. The victim, lying face down on the platform, had his head fixed to his neck by means of some soft substance, perhaps a thin sliver of clay. When a string was pulled, by moving a soldier's arm, the guillotine blade would drop to sever the head which would fall into the waiting basket. Realism was added to this macabre little display by spattering red paint around the appropriate parts of the machine. This must have been an incredibly popular party piece. These guillotines were almost gothic in their complexity, built as they were with astonishing attention to detail. Once again we have illustrated the simplest one we could find, basically because the more elaborate examples tend to detract from the main focus of attention. At the base there is a fretted fence through which protrude four small cannon at either side, and behind which are armed soldiers. There are armed soldiers at the gate which leads to a

Bone house with aviaries.

stepladder for the ascent of the prisoner. A fretted barrier either side of the ladder reaches to the high platform where the execution takes place. Here again there are soldiers in attendance, a platform for the prisoner, the high tower to house the blade, the executioner and the waiting basket. The fretted balustrade is also armed with small cannon. And, as if this were not enough, beneath the guillotine platform there is a further platform reached by a small stairway on which a fretted barrier is penetrated by more cannon behind which yet more armed soldiers stand guard. This is the most *basic* of the guillotines we examined!

A somewhat more elaborate example has at the base a cannon-defended barrier behind which there are fifteen armed soldiers. The soldiers are stationed outside a further barrier with its own gateway. The prisoner enters this inner sanctum, makes his way up an ornate stairway to a platform replete with soldiers, flagpole and flag. He is then taken up a sloping causeway to be beheaded. The most splendid guillotine of all is surely the double guillotine. This is on three tiers and is absolutely swarming with armoury and military personnel. Moreover, two 'aristos' can be despatched simultaneously.

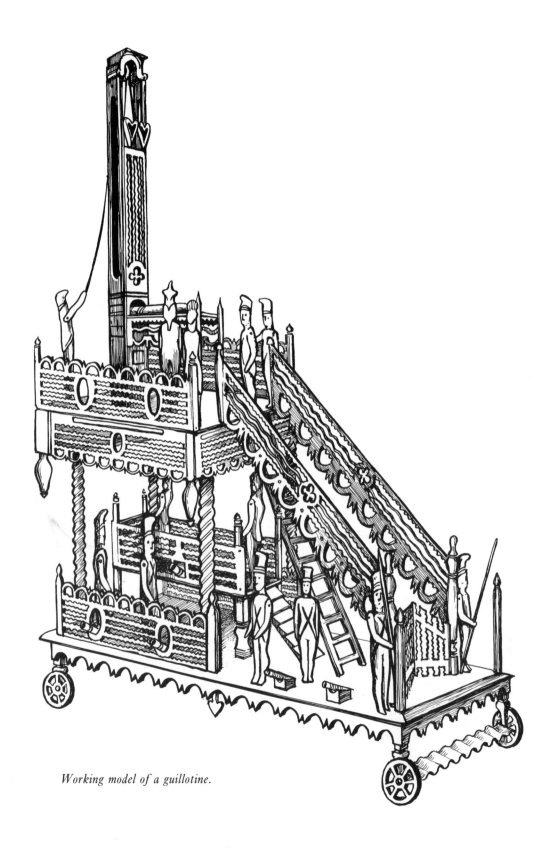

Working model of a guillotine.

But when all is said and done, it is the bone ships for which the French prisoners of war are most deservedly renowned. These models of three-masted men-o-war have pride of place in any museum which boasts possession of Napoleonic bone carvings, they are some of the most striking and beautiful of all model ships. The gleaming white lustre makes them stand out among any collection. As works of art they are unsurpassed—in any event they look right. Experts of ships are heard to complain about and even condemn these models for their inaccuracies; true, the under-water lines of these ships are not accurate, presumably because as the French did not keel-haul recalcitrant sailors, they were never able to familiarise themselves with this aspect of their ships. But the miniature guns can even be run out as they would have done in the real vessels; the bows and stern, however, tended to be

Three masted French man o' war carved out of beef bone by prisoners of war.

rather more elaborately decorated than the real ships were. But at least they looked more attractive and would most certainly have been more pleasing to the buyers. One final touch is that though all these vessels are clearly French men-o-war, they are given the names of famous British vessels of the time, and many besport the Union Jack. Perhaps they had to swallow their pride a little over this issue but as a palliative to themselves it resulted in some prisoners earning a fortune.

In the space of twenty years, thousands of imprisoned Frenchmen created a unique chapter in the history of craftsmanship and art. Through a period of adversity, they produced innumerable pieces of carving that can still delight the eye. Although their contribution was in the tradition of the mariner, these men brought it to new heights. As this work was the only outlet for their creative energies, it was channelled along virtually a single path, the only medium available being bone.

Detail of stern of bone man o' war.

9 Bones
in art

The portrayal of bones, isolated or articulated, is a subject of such vastness that it requires an entire volume to itself. Here we have an embarrassment of riches and in consequence have been forced to select a mere handful of items to illustrate our theme.

Roubiliac's monument to Lady Elizabeth Nightingale, St. Andrew's Chapel, Westminster Abbey, London, 1761.

Throughout history, and in every continent, skulls and skeletons have figured as an integral part of man's culture, as symbols of his mortality. Death is, after birth, the common experience of all human beings, although attitudes towards it vary from one society to another.

In most churches one can be fairly certain of finding a skull as a decoration of tombs. Entire skeletons also figure extensively in this role, though none perhaps quite so elegantly as that which adorns the monument to Lady Elizabeth Nightingale in St Andrew's Chapel of Westminster Abbey. The hooded and cloaked figure of Death exudes charm in spite of his sinister wielding of a spear. This life-size sculpture of a clothed and articulated skeleton is in itself a tribute to the awe and respect due to Death.

A much more macabre representation on a comparable theme is provided by Peter Bruegel the Elder in his panoramic *Triumph of Death* painted about 1562 and now housed in the Prado, Madrid.

Detail from Triumph of Death *by Peter Bruegel the Elder, 1562, now in the Prado, Madrid.*

The scene is of frenetic activity with skeletons driving cartloads of skulls and netting their reluctant victims. Although the style is different, the message conveyed is similar. Dürer, too, included skulls in many of his drawings. Frequently they appear merely as incidental ornament but in one case a single skull is emblazoned on an armorial shield.

It is not surprising that there was a preoccupation with death in medieval Europe. There was not, however, the intense concern verging on the pathological which characterised the Aztec civilisation of central America.

The Coat of Arms with a Skull, *by Dürer*.

Towards the end of the 15th century the Aztec empire became established. The Aztecs were notable for their military prowess and ruthlessness; their religion, which involved offering human hearts to their gods, was fatalistic. Emphasis was on the power of destruction, for this was a 'death culture' reflected in the items of Aztec art that survived the predations of the even more ruthless Conquistadors. One of the gods of the Aztec pantheon, Tezcatlipoca, is represented by a human skull decorated with a mosaic of turquoise. On the same theme a life-size human skull carved out of a single crystal of quartz is attributed to the Aztecs and is supposed to be of the God of Death. In spite of the sombre nature of Aztec culture, the attitude to death in Mexico, even today, is to say the least, philosophical. Death is accepted as part of the natural cycle of events, certainly nothing to evoke fear and terror. The engraving *A Jig in the Beyond* well illustrates this particular attitude. Skeletons dance, play music, drink and eat, the female ones wearing skirts to preserve their modesty!

The Mexicans, especially in their extravagant celebration of the Day of the Dead or 'Dia de los Muertos', produce a prodigious amount of skeletal ephemera such as decorated sugar skulls. These, and many other aspects of death, are discussed with great relish by Barbara Jones in her entertaining book *Design for Death*.

Mexican Art: a mask representing Quetzalcoatl or Tonatiuh, Aztec (wood), and a carved skull representing the Death God, Aztec (rock crystal).

Skeletons are indeed objects of humour and as such are one of the mainstays of the cartoonist, though there is always a hint that the laughter evoked has a degree of nervousness about it. We fear death and for the sake of our sanity, we must perforce mock it.

As well as provoking nervous laughter, the skeleton as the symbol of death is one of the most highly prized tools of the political cartoonist. We are familiar with death in such a guise stalking the battlefield—a classic for a considerable length of time. An example is a Rowlandson cartoon from about 1813 of Napoleon's Russian campaign. In the background the rival armies battle, in the foreground Napoleon contemplates his situation, while sitting opposite him a skeleton contemplates Napoleon.

A contemporary cartoonist, Gerald Scarfe, has developed a technique of distorting the features of public figures often to sickening

Napoleon on the battlefield of Leipzig by Rowlandson, 1813.

Gerald Scarfe (signature)

Gerald Scarfe's view of the Arab-Israeli war.

extremes though they remain immediately recognisable. At the same time he can achieve his effect with incredible economy of line. One of the starkest commentaries on the recent Arab-Israeli war and its futility was a simple cartoon showing two skulls in the desert. One was of a horse with nasal region bowed—the late President Nasser, the other with an eye patch—General Dayan. In many of Scarfe's commentaries on the contemporary scene the skeleton figures prominently.

A now-famous poster was produced many years ago by the Campaign for Nuclear Disarmament, as part of their publicity. It consists of two superimposed photographs—a skull and an atomic explosion. Only the C.N.D. symbol, the Nordic rune for death, identifies it as part of an organised protest movement. No comment is offered on the poster, the message is self-evident.

The Creation of Man, *by Jean Effel.*

So far, all representations of bones and skeletons have been overtly concerned with death either in religious art, politics or humour. A quite different approach is seen in the work of the French cartoonist Jean Effel. His is a gently mocking and endearing sense of humour, and bones play a key rôle in his sequence of drawings outlining the creation of man. Clearly he does not accept the Biblical account of creation—why should God have needed to rest on the seventh day if He had not needed to work?

Jean Effel shows how He must have gone about His task: beginning quite sensibly with a Meccano set of bones, He built up the framework of the body, the skeleton. These sympathetic and slightly irreligious cartoons arouse an affection for God in his white nightie. Certainly there is no hint of horror in Effel's skeletons. It is implicit that the skeleton is something perfectly normal and quite acceptable as the framework of our bodies and something that does not provoke any feelings of revulsion.

It is now accepted that the skeleton is the framework of all vertebrates and, of course, ourselves. We know about it in considerable detail and it is always something of a surprise to see the way in which the skeleton was portrayed in very early times. It seems, in fact, nothing like the skeleton as we now know it. Although death is often portrayed as an articulated skeleton, in fact when a human being rots away, the bone framework becomes very much disarticulated. In medieval times no-one bothered to look carefully at the skeleton; it was believed (and in fact one can occasionally meet people who *still* believe) that men have one less rib than women because woman was created from one of Adam's ribs. This idea, of course, was finally scotched when people began to look at skeletons. It may seem obvious now, but in medieval times one was taught not to trust one's eyes; one had to believe, one had to accept authority. It was something of a shock, therefore, when Vesalius first portrayed the skeleton as it really is. This was a great period, the early 1500s, when men began to observe Nature almost for the first time, at least for the first time in Christian civilisation. Artists became anatomists and one recalls particularly Leonardo da Vinci who studied human beings, their soft parts and their bones, with the basic motive of discovering the seat of the soul. As human corpses were not easy to come by, Leonardo supplemented his human dissections with those of other vertebrate animals. In consequence, many of his earlier anatomical drawings portray a curious mixture of human and pig organs. Only later did he come to realise that animals differed internally. Leonardo was especially concerned with the form and function of the skeleton and in his portrayals of horses and men, he recognised that form was

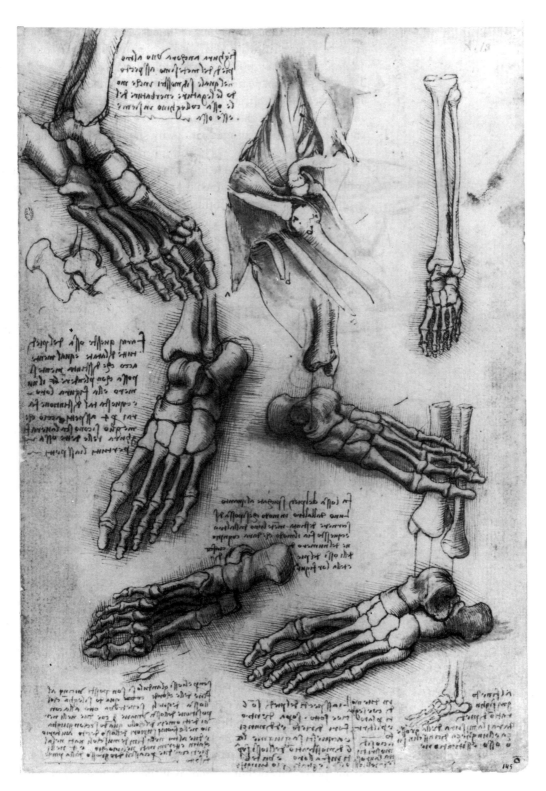

Drawings of foot bones by Leonardo da Vinci.

determined by function, by the muscles and the skeletal framework on which these muscles acted.

The modern study of anatomy began with Vesalius who made the great breakthrough in dissecting human beings and published his results in the monumental *De Humani Corporis Fabrica*. During his life-time he was patronised by royalty but his anatomical work aroused the ire of the Inquisition and Vesalius ended his life in disgrace.

Following Vesalius' and Leonardo da Vinci's time, the skeletons of many animals were studied in great detail. It was recognised that bones had a story to tell and as time went on, very strange bones were discovered, bones of gigantic animals the like of which no one had ever seen before. Skeletons held no horrors for the people who worked on these bones. Eventually it was discovered that many of the remains belonged not to animals that live today, but to animals that had died out in earlier periods of earth's history.

In the seventeenth century there were long discussions about such fossil bones. It was contended by many that these were freaks of nature, but several men believed, and as it turns out quite correctly, that they did in fact represent the remains of past life. In the last century, spurred on by the work of Darwin on natural selection and the origin of species, there was a great flowering of work on both living and extinct animals. For the latter, the scientist concerned worked almost exclusively with bones. These men were palaeontologists, and it was through their work that we learned of the existence of the giant extinct reptiles, the dinosaurs, of great sea serpents and of such things as the woolly mammoth and the sabre-tooth tiger. These are all animals with which most people are familiar, and yet no one today has seen one of them alive, we only see reconstructed pictures of them drawn by prehistoric man and modern artists. The evidence for most reconstructions relies heavily on the detailed study of the bare bones.

The giant dinosaurs seem to have a particular attraction for people who visit them in museums, particularly the giant carnivorous forms, such as *Tyrannosaurus* and its relatives. Professor Glenn Jepsen has noted that one of these forms which is displayed as a mounted skeleton in the Princeton University Museum of Natural History evokes revealing human reactions. 'Variously, people in a brief opinion-survey saw in this Jurassic reptile (1) a vigorous and determined defence of territory or of self, (2) the surging muscular tensions that precede the anticipated attack of another predator, (3) the satisfaction of a flexing yawn, (4) an over-the-shoulder snarl at an irritation, (5) a hearty guffaw, (6) the loud vocal instruction "Stop, now,—that tickles" and (7) a pleading

Poster for London Transport,
by Christopher Bradbury.

From Saga de Xam, *Paris 1967.*

shout of "Allons!" to followers.' Dinosaurs are favourite subjects for the cartoonist, and so too, are those that study them. Fossil reptiles are so familiar that they can be incorporated into the graphic arts and are perfectly acceptable. One of the finest examples we know of this sort of work is in a London Transport poster for the Natural History Museum, where the main pattern for the design is based on the skeleton of an ichthyosaur—a marine reptile with dolphin-like proportions.

Artists seem to have recognised that there is in many bones an intrinsic beauty of line and form—particularly well shown in the illustrations from *Saga de Xam*. Aspects of the human female form a complex attractive pattern, but the main centrepiece, the point to which the eye is drawn, is a human pelvis. Naturally enough it is a female pelvis, symbolic of birth.

A further illustration, now available as a poster, consists almost entirely of the female form in different positions with a single bone represented near the centre. This is the inner surface of the right temple bone of a human skull. Its shape and jagged nature contrasts astonishingly with the smooth, sinuous curves of the females.

The exact portrayal of bones certainly establishes the fact that they can be considered objects of beauty in their own right. They can indeed provide an artist with his inspiration. In the case of

Etchings of elephant skull, by Henry Moore, 1969.

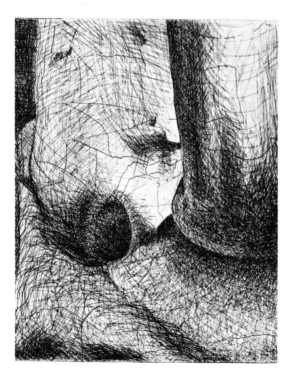
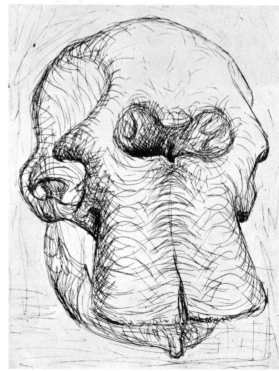

Studies and transformations of bones, *by Henry Moore, 1932 (coll. Mrs. Irina Moore).*

Skull of African elephant.

Henry Moore we know that many of his most characteristic sculptures of figures were based on the form of cow vertebrae and sheep bones. His pencil 'studies and transformations of bones' sketched in 1932, which he has kindly provided for our illustration, firmly establish the source of his inspiration.

At the time of writing, Henry Moore is currently exhibiting sculptures and etchings based on the skull of an African elephant. As he comments, 'Nature's sense of strength and structure is one of the marvellous things that you discover in studying such bones'. In a woman's magazine a photograph of the actual skull was published as an example of the artist's whimsy. It may indeed be whimsy but not that of Henry Moore! When we visited his exhibition, we could not help noticing that the actual skull was the main centre of attention. It obviously held a fascination incomparably greater that the creations of one of England's premier artists. This surely emphasises the point made by Henry Moore that bones can in themselves inspire.

It seems that in such portrayals of bones, we have come full circle—or rather we have come to the end of an evolutionary line. The initial reaction to the skeleton is that it is symbolic of death, something viewed with horror, something viewed with distaste. But as time goes on, it becomes something which is studied seriously, and out of this understanding, people are able to appreciate the skeleton as the basic frame upon which animals are built; and in this sense, it has its own beauty of line and form. Eventually the stage is reached when it is recognised that bones in themselves can be beautiful. Although it is nowadays exceedingly unfashionable to speak of progress, we think that in this instance we can say, without fear of contradiction, that the transition from a view of the skeleton as something to fill one with horror to the stage where one can view it as something beautiful and as something to provide inspiration for the artist is indeed progress.

Acknowledgements

The anatomical drawing by Leonardo da Vinci (p. 109), from the Royal Library, Windsor Castle, is reproduced by gracious permission of Her Majesty the Queen.

The authors and publishers are grateful to the following for permission to reproduce illustrations: *The Daily Mirror* for 'The Perishers' (pp. viii and 117); The Trustees of the Wellcome Institute of the History of Medicine, London (p. 2); Miss Francis Wood, Royal Dental Hospital, London and Jim Watkins, University of Reading for the radiographs of the authors (front cover and p. 3); S. K. Irtiza-Ali, Electroscan Microscope Unit, University of Reading (p. 6); The President and Council of the Royal College of Surgeons of England (p. 14); Novosti Press Agency, London (p. 64 lower); Photo. Science Museum, London (p. 97); Crown Copyright, Science Museum, London (p. 98); A. F. Kersting, London (p. 100); Museo del Prado, Madrid (p. 101); Department of Prints and Drawings, The National Gallery of Scotland, Edinburgh (p. 102); The Trustees of the British Museum, Ethnography Department, London (back cover and p. 103); Gerald Scarfe, London (p. 105); Campaign for Nuclear Disarmament (p. 106); Christopher Bradbury and London Transport (p. 111); George Proffer Publishers (pp. 112 and 113); Henry Moore, OM (pp. 114–116).

Specimens illustrated by Jennifer Middleton's line drawings come from the following collections and are reproduced by kind permission:

Ashmolean Museum, Oxford (pp. 67e, 70k, 88 left);
Arthur Bourne (p. 78 lower);
British Museum (Natural History), London (pp. 12, 49 upper, 57, 84 upper, 85 upper);
Mrs Lyn Gray (p. 64 upper);
Leo J. Heaps (pp. 17 lower, 76, 79);
Horniman Museum, London (pp. 59 a,b,c, 61, 62, 63 b,c,e,j, 65, 67 h, 69 e,g,i, 70 j,n,r, 72 upper left);
Musée de l'Homme, Paris (pp. 68 upper, 69 a,f,h, 72 upper right,

75, 77 f,g,j, 78 upper, 80 a,e,f, 81, 82, 83 lower, 87 upper);
Peterborough Museum (pp. 69c, 91 upper, 92, 93, 95, 96);
Pitt Rivers Museum, Oxford (pp. 46, 59 d,e, 60, 63 g,i, 66, 67
  b,c,d,i, 69 b,d, 70 a,b,d,e,f,g,h,l,m,o,q,s,t,u,v, 71, 73 upper,
  77 a,d,h, 80 d,g, 85 lower, 94);
Reading Museum (Duke of Wellington's Collection) (pp. 63
  f,h, 67 a);
Mr D. H. Roberts (p. 91 right);
Royal College of Surgeons of England (Hunterian Museum)
  (pp. 44, 53);
Royal College of Surgeons of England (Wellcome Pathology
  Museum) (pp. 8, 47, 48, 49 lower, 50, 51, 52);
Scott Polar Research Institute Museum, Cambridge (pp. 83
  upper, 90 lower);
Victoria and Albert Museum, London (pp. 67j, 88 lower, 89,
  90 upper).

The authors are pleased to acknowledge the generous help they
have received from E. C. Bennett, Administrative Officer, H. M.
Prison, Dartmoor; Alan Cooke, Scott Polar Research Institute,
Cambridge; Miss Jessie Dobson, Hunterian Museum, Royal
College of Surgeons of England, London; Dr A. F. Hayward,
Royal Dental Hospital, London; Leo J. Heaps; Miss Judith Levin,
Peterborough Museum; Dr Björn Kjellström, University of Lund;
Dr Marian Mlynarski, Zoological Institute, University of Krakow;
Henry Moore, OM; Dr Per Persson, University of Lund; Miss
Rosemary Powers, British Museum (Natural History), London;
Miss Rina Prentice, National Maritime Museum, Greenwich;
John G. Rhodes, Pitt Rivers Museum, Oxford; Gerald Scarfe;
Mrs Sheila Selby; John Smith, University of Reading; Dr A. J.
Sutcliffe, British Museum (Natural History), London; Professor
J. L. Turk and Sir Cecil Wakeley, Royal College of Surgeons of
England, London; Dr Calvin Wells, Norwich Museum; Mrs
Christine Wilkins.

Library facilities were provided by the Wellcome Institute of
the History of Medicine, London; Westminster City Reference
Library, London; the Royal College of Surgeons of England,
London; and the Scott Polar Research Institute Library, Cambridge.